WHY ECOSYSTEMS MATTER

WHY ECOSYSTEMS MATTER

PRESERVING THE KEY TO OUR SURVIVAL

CHRISTOPHER WILLS

OXFORD
UNIVERSITY PRESS

OXFORD
UNIVERSITY PRESS

Great Clarendon Street, Oxford, OX2 6DP,
United Kingdom

Oxford University Press is a department of the University of Oxford.
It furthers the University's objective of excellence in research, scholarship,
and education by publishing worldwide. Oxford is a registered trade mark of
Oxford University Press in the UK and in certain other countries

Published in the United States of America by Oxford University Press
198 Madison Avenue, New York, NY 10016, United States of America

British Library Cataloguing in Publication Data
Data available

Library of Congress Control Number: 2023952570

ISBN 978-0-19-288757-3

DOI: 10.1093/oso/9780192887573.001.0001

Printed and bound in the UK by
Clays Ltd, Elcograf S.p.A.

To the memories of Theodosius Dobzhansky (1900–75) and Richard Lewontin (1929–2021), and to the many other wonderful scientists who have made my own journey of discovery possible.

Acknowledgments

It is with the greatest pleasure and deepest sense of gratitude that I acknowledge the help of many colleagues on this book. Particular thanks are due to Julian Schroeder, Dick Norris, Rita Colwell, Mike McDonald, Kaustuv Roy, Rosie Redfield, Bin Wang, Qingfu Liu, Norikazu Ichihashi, Mariana Brea, Stuart Sandin, Chris Bowler, Alice Cheng, Forest Rohwer, Fang Shuai, Aspen Reese, Sara Jackrel, Lin Chao, and Andy Allen, who provided me with much support, information, advice, and encouragement. Any mistakes herein of omission or commission must of course be laid at my doorstep.

And thanks, of course, to my editors at Oxford University Press, Jamie Mortimer, Tara Werger, and especially Latha Menon. Latha has been for me, as for so many authors, an utterly reliable rock of good sense and unerring editorial judgment.

Finally, to my wife Liz, who has accompanied me on many of the adventures recounted here and many more besides, and who has always steered my absent-minded self safely through the dragons that still lurk on the fringes of the world's maps. And to my daughter Anne-Marie, who took time from her busy practice to read the whole manuscript and tighten and focus it immeasurably. Thank you both for everything!

Now, buoyed by all this love, friendship, and support, and COVID-19 permitting, I am ready to embark on new adventures. I hope to take you all with me.

Chris Wills
La Jolla, California
September 2023

Contents

List of Figures

List of Plates

Introduction

Double, double, toil and trouble;
Fire, burn, and cauldron bubble.

The witches in *Macbeth*

Evolutionary Entanglements

Almost a decade ago, my wife Liz and I embarked on one of the most amazing, and revealing, journeys of our lives. We began in the high-altitude grasslands of Peru's eastern Andes. Over the next week, covering a distance of 100 km, we descended almost 4,000 m. The vertiginous road that we followed wound down one side of a vast valley on the Andes' eastern slope.

During our descent, we encountered a rich variety of ecosystems. The Puna grasslands where we began were soon left behind, as we descended into an elfin or small-tree forest populated by small subalpine trees with twisted limbs. The tiny trees were replaced in turn by a rich and dense cloud forest, wrapped in mists. We emerged from their mists into an even denser foothill forest. Then, during the latter part of the trip, the terrain gradually leveled off into the vast lowland Amazon rainforest, threaded by numerous rivers nourished by Andean snows that would find their way to the Amazon itself.

The lowland rainforest landscape is a dynamic one, harboring many divergent communities. Some of these have been generated by the powerful rivers that have slowly meandered across the landscape over millennia. The meandering has resulted in a scattering of isolated oxbow lakes, left behind in the depths of the forest. These lakes form whenever the rivers break through the necks of their more extreme sinuosities. As the rivers follow these shorter paths, they leave behind the isolated sinuosities, now in the form of S- or C-shaped lakes. The lakes, like the forest, teem with exuberant life that is now free to diverge from the life in the river and in other nearby lakes.

This and other diversifications can be aided by the landscape itself. During the steepest early part of our descent, the trees, birds, and insects in the successive forests we passed through seemed to change dramatically with every drop of 200–300 m. A new array of colorful species of barbet, trogon, motmot, tanager, and jay appeared after every few switchbacks along the vertiginous road, only to be replaced by other, equally dazzling, species a few switchbacks later.

My initial impression of rapid change in the biological environment during our descent turned out to be surprisingly accurate. Later I found a report of a survey that had been carried out along this same road by an international collaboration of scientists, a year before our trip [1]. The survey showed that with each 250-m change in altitude, 50 percent of the bird species and an astonishing 75 percent of the tree species were replaced with other species of birds or trees. Clearly, the living worlds that inhabit successive layers of the eastern slopes of the tropical Andes are some of the richest and most dynamic on the planet. These ecosystems form the gateway to the lowland forests, where different landscape features generate yet more kinds of diversity.

Why do these mixes of tree and bird species change so much with small variations in altitude? The temperature, rainfall, humidity, and amount of sunshine differ very little across such small height differences. Indeed, when the authors tried to fit predictive models to their data, the models that only followed the effect of changes in altitude performed poorly at predicting bird species. More complicated models, especially those that included the composition and abundance of the different species of tree in an area, were much more successful at predicting which bird species would be present.

The survey's results showed that the mix of tree species is important in predicting which bird species are associated with them. But saying that the mix of A *predicts* the mix of B is not the same as saying that the mix of A *determines* the mix of B. In fact, there is ample evidence that interactions between trees and birds are not simply the result of a one-way process by which the species of trees that are present determine which species of bird live in them.

During our trip, we encountered a vivid demonstration of a reverse interaction, in which birds influence trees. Early one morning we visited a densely wooded lek. The word *lek*, Swedish for "playground," has been borrowed to serve as the scientific term for a site at which males display to attract females. In this densely forested lek, several blazingly scarlet male Andean cocks-of-the-rock were displaying.

Female cocks-of-the rock come to the lek, attracted by the males but also intrigued by the lek's other offerings. Studies have shown that, after decades of courting and mating activity by cocks-of-the-rock at these leks, the nature of the leks' tree populations can actually change. Over the years, both the displaying males and the females that they attract tend to deposit seeds from meals of their favorite fruits. The resulting trees make the lek even more alluring [2].

The trees are the precise equivalent of the fast-food restaurants that line the main streets of towns in the California Central Valley, the function of which was immortalized in the film *American Graffiti* (1973). The roadside restaurants have sprung up to nourish both the boys who cruise up and down in their fast cars and the girls whom they hope to attract. At the leks, of course, the trees have grown up simply as a digestive-system consequence of the birds spending time there. But in both cases, the end result is the same.

Other interactions between trees and animals abound in forests, and can also have profound effects on the species involved. At the end of our steep descent of the Andes, in the rainforest of the great alluvial Amazon basin itself, we encountered even more complex *evolutionary entanglements*.

I hasten to point out that the evolutionary entanglements that are explored throughout this book are equivalent to those found in the "entangled banks" described by Charles Darwin at the end of *The Origin of Species*. They have nothing to do with the bizarre *quantum entanglements* of subatomic particles.

Species caught up in evolutionary entanglements are continually evolving as they interact. Each species in an entanglement influences the characteristics of the other species—sometimes many other species—with which it is entangled, and vice versa.

Near the end of our descent, we approached another of these entanglements with caution. On the banks of rivers that cut through the rainforest, and in other recently disturbed rainforest areas, species of tree that are able to grow quickly can take advantage of brief windows of abundant sunlight. *Cecropia* trees, with their striking umbrella-like leaves, are among the most effective of these opportunists. Given open sunlight and room to grow, they are able to increase in height by a jaw-dropping 3 m in a single year.

Cecropia trees grow too quickly to accumulate much in the way of herbivore-repelling chemicals. They rely instead on helper ants of the genus *Azteca*. These big-headed ants live inside the hollow stems of the trees. They attack the many other insects, and even the mammalian browsers, that have the temerity to try to nibble on the tender *Cecropia* leaves.

The ants are fierce—hence our caution as we peered at them scurrying about. Indeed, ecologists who approach the trees too closely soon find themselves doing what might be called the *Azteca* shimmy. This dance, examples of which may be enjoyed on the web, is entertaining for onlookers but not for the painfully bitten participants.

Biologists have termed such reciprocally beneficial interactions between species *symbiotic mutualisms*—living together in a mutually beneficial way. But a closer examination of this mutualism quickly reveals it to be more complex than a two-species interaction.

The *Cecropia* trees have evolved elaborate ways of providing for their *Azteca* guests. The trees are genetically programmed to develop thin places in the walls of their jointed stems, positioned so that the ants can easily chew small holes and enter the hollow chambers that are inside. It is here that the ants set up housekeeping. Other genetically programmed welcoming features include special structures at the bases of the leaves of the host trees that excrete carbohydrate-rich packets of food called Müllerian bodies on which the ants can dine.

In return for this food and shelter, the *Azteca* ants defend against marauders, especially leaf-cutter *Atta* ants and their relatives. As workers of these ants, which are common throughout much of the American tropics, cut pieces out of leaves and carry them to underground chambers in their nest, another specialized caste of the ants cleans the pieces carefully. Once there, the carefully arranged leaves serve as the food source for edible fungi that the ants nurture and use as their main food source. Given the opportunity, and in the absence of the ferocious *Azteca*, these leaf-cutters would quickly denude the otherwise defenseless *Cecropia* trees.

The *Azteca* also attack many other browsers, responding swiftly to chemical signals from the *Cecropia* that it is being damaged. The plants have coevolved with them, and can aid them to hold down even the largest attacking insects. Hooks on their legs of the ant defenders are able to interlock, Velcro-like, with specialized hairs on the *Cecropia* leaves.

The *Azteca* fight invading plants as well as animals. They have been observed to attack slender young vines when they find them growing on the *Cecropia* branches. The vines are a serious threat because they can grow quickly, weighing down and strangling the delicate *Cecropia* saplings.

The ants, working together, chew at the vines and destroy their conductive tissues, causing them to wither and drop away [3]. These multitasking ants even keep their host plants free of any debris, so that their host's leaves can photosynthesize more efficiently.

The more researchers peer into this evolutionary entanglement, the more complex it becomes. It turns out not to be a "pure" mutualism. The *Azteca* ants quite literally milk the relationship with their host *Cecropia* trees for all that they can get. They do not repel all the invaders of the trees. On the contrary, the ants encourage scale insects and aphids to suck sap from the host tree. The ants then lap up the still-nourishing droplets of partially digested sap that their "domesticated" sucking insects excrete [4].

This *Cecropia–Azteca* mutualism provides ecological opportunities for a wide array of species like these scale insects and aphids, niches that would not exist if the *Azteca* and *Cecropia* were to lead separate lives.

The *Azteca–Cecropia* mutualism is widespread in lowland rainforests, especially in regions where there are many leaf-cutter ant colonies. It is much less common or nonexistent among *Cecropia* species that grow at higher altitudes and on islands of the Caribbean where leaf-cutter ants are rare or absent [5]. In these regions, in the absence of the mutualism, the *Cecropia* species also do not have such ant-friendly features as thin cell walls and Müllerian bodies.

In effect, the *Cecropia*–ant mutualism provides the *Cecropia* trees with a kind of immune system, defending them against the wide range of plant and animal herbivores in the teeming lowland rainforests. The ants that confer this "immunity" from predators are, of course, very different from our own immune system, which is built into us.

The *Azteca* ants are distinct from other, much more limited, defense mechanisms that the *Cecropia* trees themselves possess. The ants are a separate species, with their own set of genetic information, so that they are able to evolve on their own. Their range of behaviors can evolve as the nature of their mutualistic interaction changes with time.

This ant–tree interaction is beneficial but not essential. Even in the teeming world of the rainforest, both species could probably survive without the other, though they would do so less successfully.

The interaction is also delicately balanced. The ants cannot take too great an advantage of their host plants, for example, by raising too many sucking insects on them, without destroying the host itself. The trees cannot save energy, perhaps by cutting down on the number of energy-requiring Müllerian bodies that they produce, without decreasing the numbers of their protective Praetorian guard of ants. The innumerable benefits and constraints that maintain this interlocking mutualism are dizzying to contemplate.

Evolutionary Cauldrons Generate Evolutionary Entanglements

We returned from our trip overwhelmed by what we had seen. I had already spent some years, in collaboration with forest ecologists from around the world, exploring the ways in which tree species that share a forest influence each other's growth and reproduction. Ecologists had shown that some of these influences are the result of interactions between the trees and their fungal pathogens, and their insect and animal herbivores. These interactions help to maintain a great diversity of tree species in the forests.

The Andean trip had provided multiple examples of just how complex such interactions can be, and the pre-eminent role that such interactions play in the evolution of ecosystems. In this book, we will explore how our growing understanding of such complexities has begun to change how we see the living world.

In the course of my forest research, I have been able to travel widely across six continents and to visit many island archipelagos. I have seized these opportunities to explore not only remote forest ecosystems but also a wide range of other terrestrial and marine ecosystems. My work as a population geneticist and evolutionary biologist has enabled me to set stories of these travels in an evolutionary context [6].

This book goes further, drawing on the amazing work of thousands of scientists who have been empowered by the revolutionary discoveries of molecular biology to probe the Earth's ecosystems in unparalleled genetic detail. I will use their findings to suggest some answers to the existential questions that confront us as we come to terms with the reality that, in order to survive, we must become stewards of the planet's health and sustainability.

Our Andean trip was a vivid reminder of how many wonderful, and sometimes essential, complexities of the living world we are in danger of losing. How can we ensure the long-term survival of the complex, rapidly evolving ecosystems on which the world's living creatures depend? And what properties of these ecosystems are essential for their (and our) long-term survival?

We will see how some remarkably complex ecological interactions can evolve swiftly, even during the course of laboratory experiments that start with genetically uniform populations and last for weeks or months. We will explore emerging technologies that are being used to characterize and quantify these ecological interactions, along with the evolutionary changes that gave rise to them. Answers to these questions are sometimes unexpected and always amazing.

These experimental approaches are showing us how quickly and how easily evolution can generate ecological complexity. And they point us to an astounding conclusion: every one of Earth's teeming ecosystems is a kind of evolutionary cauldron. The cauldrons are all bubbling with new mutational changes. Some of these changes can take an entire ecosystem—and occasionally the entire planet—in an utterly new direction.

We can only imagine the ghastly potion that was brewing in the witches' cauldron in *Macbeth*. Evolutionary cauldrons are much less frightening. The witches could have ruined their mix by adding the wrong ingredient. Evolutionary cauldrons have the capacity to fix things that go wrong, by providing many routes by which new mutations in their component species can express their beneficial qualities and by forcing evolutionary changes in the properties of disruptive invading species that will minimize the invaders' impact.

W. James Kent and colleagues in 2003 may have been the first to describe evolutionary cauldrons [7]. They used the term to describe the rapid evolutionary changes that are possible in both mice and humans because of the many highly mutable short repeated regions that both species carry in their DNA. Here, I use the term far more broadly, to encompass the entire ferment of evolutionary changes that emerge in all ecosystems as their species interact with a bewildering variety of physical and biological challenges.

New directions in evolution are not rare. As we will see, they are more likely to happen in the world of very small organisms. But they can take place anywhere and at any time. Evidence is accumulating that the cauldrons most likely to produce such evolutionary ferment are the ones that are undamaged and brimming with diversity.

It will be essential, as we try to stabilize the ecosystems on which our planet depends, that we do not limit their ability to continue evolving. To do so would limit the ability of the species in these ecosystems to adapt to the unavoidable changes that the future holds.

We will see that healthy evolutionary cauldrons are also protective. Their very diversity can shield their component species from runaway diseases and other destabilizing events.

Ecologists of the past observed interactions between species of organisms that are visible to the naked eye. Occasionally, they ventured into the world of the small creatures that are visible in the light microscope. But we will venture further, to discover how interactions that involve species drawn from all of life's levels play a role in every ecosystem. Organisms at each level of size and complexity, from viruses to whales, inhabiting ecosystems from the deep sea to the most extreme of Earth's deserts, play important roles in keeping evolutionary cauldrons active.

We will trace in detail how and why the living world of today has evolved to be far more complex, and may actually be even more resilient, than the worlds of the past. We will see how the process of evolution itself, nurtured by the complexity of the living world that it has produced, has also changed. Its mechanisms have not changed—it is still Darwinian evolution. But as ecosystems have increased in complexity, the opportunities for the species involved have increased in number. Evolution now has more ways of bringing about genetic change than it did in the past, both in the world of microbes and in the world of more complex multicellular organisms—including ourselves.

We will also look at why some ecosystems are more vulnerable than others. We will join the search for the most likely interventions that can stabilize, and even undo, the damage to such unusually endangered ecosystems.

These new insights give us great power, but they also confront us with a choice. We can continue on our present course and simply extinguish these remarkable evolutionary cauldrons that nurture rapidly evolving, interacting swarms of species. We have tried that approach for tens of thousands of years, and it has led us to the brink of devastating planet-wide mass extinctions. Alternatively, we can try to understand the cauldrons' sometimes quite unexpected capabilities, and use that understanding to help ecosystems recover from damage. We have followed the first course to the edge of disaster. It is time to try the second.

I

How Darwin Brought Evolution and Ecology Together

It is interesting to contemplate an entangled bank, clothed with many plants of many kinds, with birds singing on the bushes, with various insects flitting about, and with worms crawling through the damp earth, and to reflect that these elaborately constructed forms, so different from each other, and dependent on each other in so complex a manner, have all been produced by laws acting around us.

Charles Darwin, *The Origin of Species*, 1st edition (1859), first sentence of final paragraph

In this famous passage at the end of his most famous book, Darwin laid down a vast challenge for himself and for the scientists who would follow him. He chose, as an example of the power of evolution through natural selection, an "entangled bank," a complex evolutionary entanglement of organisms all living together in what would come to be called an ecosystem. He asked his readers to reflect that this complexity had been produced by the operation of natural selection and of other natural processes.

Darwin also realized that scientists must do more than reflect on such a scenario. He knew that it would be up to his successors to prove him right or wrong about the processes that had driven such entanglements, and to explore their full implications.

The actual low wooded escarpment, known as a "bank," that Darwin used in his illustration of evolution in action seems now to have been identified. Clues in Darwin's papers suggest that it may be a ridge, home to undisturbed stretches of woodland and meadow, that Darwin could easily have reached along footpaths from his house in Kent.

This little ecosystem is the Orchis Bank, named after the wild orchids that abound there. It has now been proposed as a World Heritage Site.

In Darwin's time, these vegetation-covered banks, along with other bits of wildland such as the hedgerows that for centuries had marked the boundaries between tiny plots of medieval farmland, could be found everywhere in the English countryside. For more than a millennium, since the fields were first cleared, these little areas have provided refuges for wild creatures.

Twenty years ago, my uncle Norman Done and I were standing near a hedgerow that bounded a field in Northamptonshire. We watched the riders of the Pytchley Hunt, established nearby in the late seventeenth century, as they galloped away in pursuit of a fox. Their hunting horns sounded across the peaceful countryside, and their hounds bayed in counterpoint. The horses, with their red-coated riders, leapt over a far hedge and disappeared.

Moments later, no more than twenty feet away, the object of all this pageantry poked his (or her) head out from a nearby clump of bushes. We and the fox regarded each other for a long moment, and then the fox also disappeared, followed by our good wishes.

This is a moment I will always treasure, because it encapsulates the ecological wonders that hedgerows still protect. Today, alas, the remaining banks and hedgerows are merely irritating impediments to the clear sweeps of denuded countryside that are being created to meet the demands of modern farming machinery.

Darwin asked his readers to imagine the amazing processes that had shaped such beautiful but otherwise unremarkable features of the bucolic English countryside of the time. He could have raised the ante. From his travels on the *Beagle*, he was familiar with ecosystems that were far more spectacular.

He could have used the teeming rainforest of Bahia in Brazil, through which he had roamed in dazed wonder near the beginning of his voyage. Or he might have asked his readers to visualize the teeming coral reefs that surround the atolls of the Keeling Archipelago to the west of Australia. He had explored those reefs at one of the *Beagle*'s last ports of call, and had used his observations to support his theory of coral atoll formation.

Such examples might have dazzled his readers, but they might also have puzzled them. Their experience of such things would have been limited. Darwin knew his readership, and he knew that the tree-covered bank so near his house was representative of ecosystems with which they would all have been familiar.

Since Darwin's day, the bank example in the *Origin* has been drawn on many times by evolutionists and ecologists. It has been pressed into service as a jumping-off place for a number of evolutionary theories.

The most well-known of these theories is the "tangled bank" hypothesis of Michael Ghiselin, Graham Bell, and others [8]. Their theory suggests that intense competition for resources among the many species in complex ecosystems will favor sexual over asexual reproduction. This is because the diverse offspring produced by a sexually reproducing species are more likely to include some that are capable of surviving in such a crowded and varied environment than are the genetically uniform offspring of asexual parents.

This important aspect of an entangled bank has been explored elsewhere. But here we will examine recent progress in answering Darwin's original challenge, which was to understand fully how the bank and all other ecosystems have evolved into the states that we currently observe.

How entangled are the species in such a bank? How have the evolutionary changes in individual species resulted in, and been shaped by, evolutionary entanglements with other species in the bank? What role do these entanglements play in maintaining the integrity of the entire ecosystem and in shaping its evolution? Ultimately, how essential are these various entanglements to the survival of the living world?

To understand Darwin's entangled banks, we must find ways to build on the simpler process of natural selection acting on individuals—which is the primary way by which evolution takes place. As we learn more about ecosystems, we are discovering many routes by which selection-driven evolutionary changes taking place in other species can alter a particular species' evolutionary trajectory.

It is here that the implications of Darwin's language about his entangled bank become particularly telling. Note that Darwin had, at least initially, written *entangled*, not *tangled*.

To get down into the grammatical weeds for a moment, we find that *tangled* is a descriptive participial adjective, derived from the verb *to tangle*. It suggests simply that the organisms are all tangled in some fashion with each other. But *entangled*, which is also a descriptive participial adjective, is derived from the verb *to entangle*. This adjective goes further, in that it encompasses the possibility that the organisms might be actively participating in the entanglement process.

As Darwin made revisions for the fifth and penultimate edition of the *Origin* (1869), his normally superb sense of precision in the use of language seems to have deserted him for a moment. Perhaps, in order to tidy up the paragraph a bit, he substituted the shorter "tangled" for "entangled." I suggest this edit was made in error. Clearly, Darwin's banks are actively entangled. And entangled they remain.

We glimpsed some of these evolutionary entanglements in the Introduction, when we encountered the amazing pseudo-immune system provided by the ant–plant mutualism that has evolved to protect lowland *Cecropia* trees. We now know that such entanglements extend to every level of complexity, and may involve the simultaneous interactions of hundreds or even thousands of species.

Evolutionary entanglements can lead to explosions in diversity. As we will see in this chapter, Darwin glimpsed a hint of this process in the Galápagos Islands. But it is only now that we can begin to measure and understand their true importance to the living world. And thereby hangs a tale—of discovery, and perhaps of salvation.

The *Beagle* and the Bank

The journey down the eastern slope of the Andes that began this book had brought vividly home to me just how entangled the natural world can be. But it was not until several years had passed, and amazing new findings from around the world had further illuminated these evolutionary entanglements, that I began to comprehend fully how actively entangled the field of ecology and my own fields of genetics and evolutionary biology really are.

The path that led Darwin to his entangled bank is also a story of his own continually widening scientific horizons. As in all such scientific journeys of discovery, he had to discard entrenched and seemingly obvious ideas in favor of newer concepts that were a better fit to the facts and that might be able to survive experimental tests.

The tiny Royal Navy brig HMS *Beagle* arrived in the remote Galápagos Islands, 1,000 km to the west of the Ecuadorian coast, on September 15, 1835. The *Beagle* was nearing the end of its detailed survey of South America's dangerous coastlines, a survey that had been made possible because of its precious cargo of twenty-four chronometers.

These chronometers, newly invented precision timepieces, could maintain their accuracy even in rough seas. They had all been set to noon at the Greenwich Observatory at the beginning of the voyage. They had enabled the *Beagle*'s navigators, by comparing local noon to Greenwich noon, to map the precise longitude of the innumerable rocks, reefs, and unexpected islands that abounded around Cape Horn, the most perilous region in the world for the sailing ships of the time.

The *Beagle* was carrying another precious cargo, though just how valuable was not yet apparent. This was the twenty-six-year-old Charles Darwin, fresh from his studies at Cambridge. Although he was listed in the *Beagle*'s manifest as a gentleman companion to the ship's commander Robert FitzRoy, unofficially he was the ship's naturalist.

Over a brief six weeks in the archipelago, Darwin had the opportunity to go ashore on four different islands. He worked frantically, scrambling over the broiling-hot, razor-sharp rocks to survey the geology, take samples of the plants, reptiles, birds, and mammals, and absorb everything that he could find out about the islands. The end of each day was spent trying to fit what he was learning into the mass of information that he had already accumulated in the more than three and a half years since the voyage had begun.

Darwin had been strongly recommended by his Cambridge botany professor, John Henslow, as a companion to FitzRoy. But his father Robert Darwin, a medical doctor, initially forbade him to go. Luckily for science, Dr. Darwin was persuaded by his own brother-in-law Josiah Wedgwood, the son of the founder of the Wedgwood pottery firm, to change his mind. Wedgwood perceptively described his young nephew as "a man of enlarged curiosity," and pointed out that the voyage would afford him "such an opportunity of seeing men and things as happens to few."

Now, Charles was amply confirming Wedgwood's assessment, as he absorbed and began to comprehend the principles that underlay the deluge of observations he was making about creatures and phenomena that few Europeans had seen or experienced.

His enlarged curiosity, which had so impressed his uncle, allowed him to break free from the patterns of thought about the natural world that had circumscribed even the most brilliant thinkers of his age. But the process was not easy. When the voyage began, he still thought about nature from the viewpoint of the theologian and philosopher William Paley, whose *Principles of Moral and Political Philosophy* had formed an important part of Charles' curriculum of study at Cambridge.

Paley viewed the natural world as a divinely ordered mechanism. The living world's very complexity, and the elegance with which it functioned, were indisputable evidence that it had been designed by the Creator. To Darwin (and to many others), this made perfect sense.

Paley's ideas were appealing in other ways, too. As he was growing up, young Charles had been repeatedly exposed to the strong anti-slavery views that had been held by his grandfather Josiah Wedgwood and that were now a bedrock belief of Wedgwood's entire family.

Paley's *Principles*, which as a student Charles had to read and explore in great detail as part of his curriculum of study at Cambridge, included devastating arguments against slavery. Paley saw slavery as an appalling exception to the Creator's elegant order. His arguments played an important role in shaping England's anti-slavery movement.

When young Charles actually encountered the savage reality of slavery in Brazil and Argentina, it had a great impact on him and must have made him even more sympathetic to Paley's ideas.

Darwin's journals of the voyage show that he attempted to fit his observations and discoveries about the natural world into Paley's beautiful scheme. But, as we will see, he kept encountering cases in which organisms did things to each other that a kindly Creator ought never to have permitted.

Further, species were not unchanging, as Paley's scheme had assumed, and that formed a central tenet of the entire Church establishment. In Patagonia, Darwin had found fossils clearly related to, but nonetheless distinctly different from, animals that were living in the area during his visit. He became more and more convinced that species must have come and gone. The real world, it seemed, was far from an elegantly designed, finely balanced mechanism that preserved the original creation in detail.

There were also tantalizing patterns to the ways in which species had appeared and disappeared. As the *Beagle* was slowly working its way north, Darwin had an opportunity to travel across the Andes from west to east and then back (at the latitude of Santiago, which lies far to the south of the roistering tropical world that my wife and I had encountered). He saw that the plant and animal species on the western and eastern slopes of the mountain range were similar but not identical. Why?

The Andes themselves provided him with an explanation. While he was ashore near the town of Valdivia on Chile's southern coast, he experienced an earthquake so strong that it threw him to the ground. Later, a survey by FitzRoy and members of the *Beagle*'s crew found that beds of intertidal mussels along the coast had been thrust ten feet into the air by the earthquake, so that they were now gaping and rotting in the sun.

Darwin, well-prepared by his careful reading of Charles Lyell's *Principles of Geology* (1830–3), began to wonder whether the Andes themselves had been formed by a long history of such land-raising earthquakes. The implications, he realized, were vast. When he traveled across the Andes and back, he observed similar but divergent groups of species on their eastern and western slopes. Perhaps these species were descendants of more uniformly distributed plants and animals that had lived in the region as the mountains were just beginning to form. The differences in the species on the slopes might, he eventually realized, be driven by different selective pressures as the species became separated by the raising of the land between them.

He was further reinforced in these conclusions by what he could see of the mountains' amazing geology. As he was crossing the immense Andean range for the first time, at an altitude of 2,500 m, he encountered a petrified forest. The trees had been suddenly buried and frozen in time by great deposits of volcanic ash. This ancient catastrophe had recently been revealed by recent erosion of the softer ash, leaving the still-vertical tree trunks exposed. The forest, Darwin concluded, must have been growing under warm and wet conditions, far different from the cold, dry, and windy world of the mountain slopes of his own time.

As he explored further up the slope that lay behind the petrified trees, he encountered layers of sedimentary rock that gave him the impression that the volcanic deposit containing the ancient forest had been overlaid with a thick layer of marine sediments. This led him to conclude, wrongly, that the forest had first been entombed by ash, then submerged beneath an ocean, and finally raised to its present height.

Darwin was not sure of the forest's age, but he guessed it had thrived at some point early in our current Age of Mammals. We now know that it is much older, chiefly made up of seed ferns and evergreens that were typical of the Triassic Period about 225 million years ago. It was likely home to some of the earliest dinosaurs, and to other reptiles that had some mammalian characteristics. Some of those mammal-like lineages would evolve into the mammals.

Darwin had indeed found evidence of a great upheaval, but it was not quite as dramatic as he had hypothesized. Mariana Brea, of Argentina's Universidad Naçional de la Plata, and her colleagues have studied the forest and its surrounding sediments carefully. They have found no sign of marine layers [9]. The overlying sediments that Darwin had mistaken for marine were actually deposits left behind by winding rivers flowing into estuaries and lagoons. The forest had certainly remained intact and safely entombed through many great geological events—but it had not, it appears, been plunged into the depths of the sea and then raised again.

We now know that the great mountain range of the Andes, the world's second highest after the Himalayas, was formed as the result of a slow, grinding encounter between two of the Earth's tectonic plates. The Andes' peaks and high valleys are the immense crumpled result of a vast slow-motion collision between the eastern margin of the oceanic Pacific tectonic plate and the western margin of the much thicker continental South American plate.

The Pacific plate is being forced under the South American plate, thrusting it up. This collision has lasted for sixty million years, and it continues today. The resulting pileup has pushed different layers of rock thousands of meters into the air. And some of these rocks did indeed come from beneath the ocean. During his journey, Darwin saw, and recognized, that some of the higher-altitude strata contained ancient marine fossils.

These and other adventures had already set Darwin's mind ablaze before the *Beagle* reached the Galápagos. His visit to the islands, though important, was actually only a small part of his journey of discovery.

The brevity of the stay in the islands, combined with the difficulty of scrambling across the forbidding landscape in the blazing sun, made Darwin's surveys challenging. Luckily, he brought essential skills to the task. During a brief (and quickly abandoned) stint as a beginning medical student at the University of Edinburgh, he had taken lessons in the rudiments of taxidermy. This enabled him to preserve a number of the small insect- and seed-eating birds that would later become known as Darwin's finches. FitzRoy and several crew members also collected the birds.

Unfortunately, Darwin accidentally mixed up the few specimens that he had time to obtain, so that he could not tell which islands they had come from. And, because he was a novice at distinguishing species, he initially misidentified some of the finches as wrens or as relatives of the orioles.

Mixed up or not, the finches fascinated him. He was already beginning to think that radiations of distinct species from a few founding ancestors must have happened on the different islands of the Galápagos—just as might have happened on a vaster scale on the eastern and western slopes of the Andes. The finches might be examples of such divergence.

Sorting out the finches and seeing what patterns they showed had to await Darwin's return to England. But during his Galápagos visit, he was already seeing hints of such radiations in other organisms. One striking case involved three closely related species of mockingbird that inhabited different islands. All were distinctly different from each other and from the mockingbirds on South America's west coast.

Darwin was able to send some of his collections back to England on ships that the *Beagle* encountered during its voyage. By the time the *Beagle* returned to England, five years after its departure, experts in many fields were already at work on Darwin's samples.

Luckily, he had collected the plants of the Galápagos more methodically than he had the finches. His Cambridge mentor John Henslow confirmed that several of the closely related plant species in Darwin's collections were indeed from different islands. And the ornithologist John Gould was able to obtain hints that the finches showed similar patterns.

After his return to England, stimulated by these findings, Darwin began the first of a series of notebooks on the transmutation of species. If species were in fact becoming more different with time on different islands of the Galápagos, and perhaps also on the western and eastern slopes of the Andes, what was the cause? He soon came to the realization that a process of natural selection could account for much of what he had seen.

In his autobiography, published at the end of his life, he recounts how, shortly after his return to England, his vague ideas were crystallized into a real working hypothesis by his reading of Thomas Robert Malthus' *An Essay on the Principle of Population* (1798). Malthus had pointed out that human populations increase exponentially in numbers—that is, they tend to increase by some proportion, not by some constant number, each generation.

The effect is that human populations increase in total size more and more quickly with time. Even if their resources also increase, those increases are unlikely to follow similar exponential curves. Disaster will be the inevitable result. Exploding human populations will be limited when they run out of their resources, mostly in unpleasant ways.

Darwin realized that this principle should apply to all populations of living things. And he went much further than Malthus. He saw that the members of all of these exponentially growing species that had the greatest ability to compete for their limited resources would be the ones with the greatest chance of surviving and reproducing.

Darwin realized that organisms other than humans, though unable deliberately to change the terms of the competition for resources, have another recourse—natural selection. To the degree that the capability for survival of each organism is heritable, this heritable component will influence whether or not those characteristics will be passed to subsequent generations. This realization was the great insight that would eventually lead him to the entangled bank.

In 1842 and 1844, he clarified his thoughts by writing essays that explored the new theory and the many things that it could explain. The edits and emendations that Darwin made to his essays as he wrote them were included in the transcriptions that were published by his son Francis in 1909 [10].

The essays showed that he proceeded cautiously. For example, in the 1842 essay, he reminded himself: "Ought to state the opinion of the immutability of species and the creation by so many separate acts of will of the Creator."

Additionally, in the 1842 essay, Darwin tried to have it both ways. He suggested that the often-cruel operation of the process of natural selection should, because of the wonders that it has produced, enhance our awe at the powers of an omniscient God. God had created a world in which natural selection could produce beautiful things.

As Darwin must have known, he was violating the principle of Ockham's Razor, which is that "entities (that is, explanations) should not be multiplied beyond necessity." God could have produced the living world, and so could natural selection, and indeed both could have played a role, but there was no *necessity* for them both.

He probably also recalled the polymath Pierre-Simon Laplace's famous (but possibly apocryphal) response to Napoleon's query about whether Laplace believed in God: *"Je n'avais pas besoin de cette hypothèse-là!"*

I like to translate this in a way that takes into account the emphatic suffix "-là!": "I have no need of that *particular* hypothesis!"

But Darwin was constrained by the beliefs of his family and his immediate society. He was also seeking patterns in an immense body of biological and geological information, much of which was being supplied by others. He depended on interactions and collaborations with dozens of colleagues in England's scientific establishment, and on contacts with thousands of correspondents from around the world. Their friendship and cooperation were essential as he began to explore the implications of his theory. He could not afford a reputation as a godless iconoclast—though of course he would soon turn out to be one of history's greatest overthrowers of idols.

Even as Darwin was trying to reconcile theology and natural selection, he was already following up on many of natural selection's biological implications. In subsequent years, he would explore the evolution of our own species, the major role that is played by sexual selection in evolution, and the vast evolutionary implications of sexual reproduction and outbreeding.

He realized that species with elaborate adaptations should be expected to have close relatives that show similar but less extreme adaptations. He used this "principle of continuity" with a species' relatives to cast light on the evolution of insectivorous and climbing plants, the expression of emotions in the mammals, and the development of secondary sexual characteristics, along with many other natural phenomena.

He was also pushing the explanatory powers of natural selection to the limits of what was known—and beyond. One of these boundary-pushing ideas forms the starting point of this book.

The Origins of the Entangled Bank

We have briefly traced how Darwin had reached the idea of natural selection. But there were still many gaps that needed to be closed in the journey from his original powerful idea to its full realization in the form of a deep understanding of how entangled ecosystems have evolved. Darwin was able to close some of these gaps, but many others remained.

In his 1845 account of the voyage of the *Beagle*, Darwin shows us how he came to appreciate the evolutionary implications of species divergences among the different islands of the Galápagos:

My attention was first called to this fact by the Vice-Governor [sic], Mr. Lawson, declaring that the tortoises differed from [sic] the different islands, and that he could with certainty tell from which island any one was brought. I did not for some time pay sufficient attention to this statement, and I had already partially mingled together the collections from two of the islands. I never dreamed that islands, about fifty or sixty miles apart, and most of them in sight of each other, formed of precisely the same rocks, placed under a quite similar climate, rising to a nearly equal height, would have been differently tenanted.

Later, in the *Origin* itself, he expanded on his and Lawson's observations to present a remarkable prediction about the nature of the selective pressures that drive such divergences. He also laid out how difficult it had been to reach this conclusion. He had been forced by his data to confront, and then to abandon, an ingrained misconception about how natural selection ought to work. It was a challenge for him to overcome what he called a "deeply-seated error." I have italicized substantial parts of this passage to add emphasis.

[T]he several islands of the Galápagos Archipelago are tenanted, as I have elsewhere shown, in a quite marvellous manner, by very closely related species; so that the inhabitants of each separate island, though mostly distinct, are related in an incomparably closer degree to each other than to the inhabitants of any other part of the world. And this is just what might have been expected on my view, for the islands are situated so near each other that they would almost certainly receive immigrants from the same original source, or from each other. But this dissimilarity between the endemic inhabitants of the islands may be used *as an argument against my views*; for it may be asked, *how has it happened in the several islands situated within sight of each other, having the same geological nature, the same height, climate, etc., that many of the immigrants should have been differently modified, though only in a small degree. This long appeared to me a great difficulty: but it arises in chief part from the deeply-seated error of considering the physical conditions of a country as the most important for its inhabitants; whereas it cannot, I think, be disputed that the nature of the other inhabitants, with which each has to compete, is at least as important, and generally a far more important element of success.* [...] *[A] plant, for instance, would find the best-fitted ground more perfectly occupied by distinct plants in one island than in another, and it would be exposed to the attacks of somewhat different enemies. If then it varied, natural selection would probably favour different varieties in the different islands.*

Charles Darwin, *The Origin of Species*, 1st ed., 1859, p.400.

Darwin had to abandon an assumption he had made about how natural selection ought to work, in order to build his picture of how species had spread to different islands and begun to diverge on the physically similar islands after they first arrived on the archipelago. This divergence must, he realized, have been driven by the interactions among the evolving species themselves. Without these interactions, and only a requirement to adapt to similar physical environments on the islands, they would not have diverged or would have diverged far more slowly.

He claimed that this new insight "cannot, I think, be disputed," but in fact it was far from proven. There were three great gaps that still needed to be closed in the chain of evidence.

The first of these gaps is ecological. Darwin's "deeply-seated error"—his assumption that physical factors in the environment must constitute the primary limitation to organisms' survival—is still shared by many ecologists of the present time.

The second gap is genetic. Darwin knew that these volcanic islands were young, so the adaptive changes had to be taking place relatively quickly. But it was difficult for him to see how, given the prevailing view of inheritance at the time, rapid changes could happen.

Darwin thought that most evolutionary change takes place through the inheritance of characteristics that are *acquired* by organisms during their lifetime. This old idea is often attributed— incorrectly—to an eighteenth-century proponent of evolutionary change, Jean-Baptiste Pierre Antoine de Monet, the Chevalier de Lamarck. But the idea is far older than that, going back to

Aristotle and other ancient Greeks. Lamarck codified it in his "law of use and disuse," and then embellished it further by suggesting that organisms could pass on such heritable physical changes in response to their "felt needs (*sentiments intérieurs*)." Darwin called this idea "nonsense."

In fact, Darwin's grandfather Erasmus had suggested an idea very similar to Lamarck's. Erasmus had postulated evolutionary change as a response to organisms' "own exertions as a consequence of their ideas and aversions, of their pleasures and their pains . . ." [11].

I suspect that both of these early evolutionists were trying to account for one of the large problems inherent in the idea of the inheritance of acquired characters.

Aristotle was puzzled by the glaring fact that children born of mutilated parents are themselves unmutilated [12]. I suspect that Lamarck and Erasmus Darwin, who were certainly familiar with the teachings of Aristotle, were trying to provide an explanation for Aristotle's puzzle. If organisms only passed on characters that they desired, by somehow being able to sort them out from nasty acquired characters such as missing limbs, this would solve the problem. But, if it was indeed this problem that underlaid Lamarck and Erasmus Darwin's evolutionary ideas, they did not explain clearly what they were trying to do.

Charles could not imagine a mechanism by which organisms themselves might be able to distinguish favorable from unfavorable acquired changes and pass only the favorable ones to their offspring. His proposal of natural selection, by which natural processes could sort out advantageous from disadvantageous characteristics, seemed to him to be an infinitely preferable alternative to Lamarck's and Erasmus Darwin's suggestion that organisms can control their evolution by their inner feelings.

He realized that natural selection's only requirement is that the characters it is acting on must, at least to some degree, be heritable. Even if these characters were acquired during the organism's lifetime, then natural selection, not the creatures themselves, should be able to distinguish between the favorable and unfavorable acquired characters and produce evolutionary change.

Shortly after Darwin presented his concept of natural selection, Gregor Mendel discovered the rules by which factors that influence organisms' characteristics, which we now call genes, are passed down through the generations. Mutations of a wide variety of types can change the genes. All species have the capability of drawing on, and shuffling, this mutation-generated variation in an effectively infinite number of ways. Acquired characters, such as changes in size and shape of a body part, could never have supplied such a universe of possibilities.

We now have direct proof that some mutational changes in genes can persist at low frequencies for thousands and even millions of generations [13]. Because species are possessors of huge pools of such persistent mutational variants, the random shuffling of the variants each generation can produce individuals that have new, and occasionally startlingly different, characteristics. The effects of these changes, and the exciting evolutionary trajectories that they create, are all around us. They have helped to shape the evolution of entangled banks throughout the world and throughout billions of years of history. These diverse components are simmering in all of the Earth's evolutionary cauldrons.

The Emergence of Evolutionary Ecology

The third gap that had to be closed was not yet apparent in Darwin's time. The field of evolutionary biology clearly began with Darwin and his great insight about natural selection. Like general relativity, natural selection will operate anywhere. Suppose that life exists in the vast atmospheres of gas giant planets, or in volcanically heated waters deep in ice-covered oceans of worlds that are adrift between stars. If so, natural selection will be there. It will shape the evolution of any living organisms, regardless of the details of their inheritance.

The field of ecology, in contrast, did not begin with such a thunderclap of insight. It had a much more diffuse origin. But natural selection was at its heart right from the beginning—though it has, alas, often been lost sight of since that time.

The study of populations of organisms and of their interactions was first given a name, *ecology*, in 1866 by the remarkable zoologist Ernst Haeckel. The name is derived from the Greek word *oikos* (house).

When Haeckel introduced this term, he meant it to be taken in an extremely broad sense—literally, the study of the houses of all living organisms in all their aspects, including the ways in which the members of different species are evolving through their interactions with each other:

By ecology we mean the body of knowledge concerning the economy of nature—the investigation of the total relations of the animal both to its inorganic and to its organic environment; *including, above all, its friendly and inimical relations with those animals and plants with which it comes directly or indirectly into contact*—in a word, ecology is the study of all those complex interrelations referred to by Darwin as the conditions of the struggle for existence [14]. (Translated by W.C. Allee, emphasis added.)

It is clear that Haeckel, an enthusiastic proponent of Darwinian evolution, realized that our houses, like ourselves, have chiefly been shaped by natural selection. Understanding the interactions among species must, he felt, be a chief goal of this new field of ecology.

As the field of ecology slowly emerged during the early twentieth century, however, most of the scientists who helped to shape it were not primarily evolutionists. Many of the central figures, such as Charles Elton and G.E. Hutchinson, were field ecologists, fascinated by such clearly quantifiable phenomena as interactions between predators and prey and the spread of invasive species. They were puzzled by the fact that many different species of predator and prey were sometimes able to coexist in an ecosystem.

Other workers, such as G.F. Gause and W.C. Allee, aimed to understand highly simplified ecosystems in the laboratory. Still others were mathematical ecologists, who looked at the patterns reported by workers in the field and the laboratory, and searched for mathematical equations that could predict these patterns.

Mathematical, laboratory, and field ecologists have often come together to model ecological relationships in ways that hold up well to repeated tests and that explain important aspects of the natural world. The theory of island biogeography, proposed by Robert MacArthur and E.O. Wilson in 1963 [15], and tested empirically in 1970 by Daniel Simberloff and Wilson [16], is a brilliant example. But it is, as we will see, missing something.

The theory begins by asking us to imagine an archipelago of islands of different sizes lying at various distances from a larger mainland. The mainland supports many species of animal and plant. Each of the species has mechanisms that enable its members to disperse, so that when some of them fly out to sea, or when their seeds are carried away by the wind or on the feathers of sea birds, a few will end up on the islands.

Even as these new species are arriving, other species that are already living on the islands are disappearing. A species can go "locally extinct" on an island even though it may still be present on other islands and on the mainland. Such local extinctions are especially likely if the islands are small and isolated.

The theory presumes that such chance migrations and local extinctions can go on for some time, until gain and loss in species numbers balance each other and the number of species on each island reaches a steady state or equilibrium. What variables must be known in order to predict that equilibrium number?

MacArthur and Wilson showed that the size of an island and its distance from the mainland both strongly influence equilibrium numbers. Smallness is bad for an island's diversity, because few species arrive per unit of time and the rate of local extinction is high. The problem is compounded if the small island is far from the mainland, because even fewer species will arrive and local extinctions remain high on all such small islands, regardless of where they are.

If an island is large, then on average it will have more species than a small island. But here, too, distance from the mainland plays a huge role. A large island close to the mainland will have more species than a small one at the same distance. But even among large islands, the equilibrium number of species drops precipitously with increasing distance from the mainland. Greater distance from the main source of species reduces the chance that new species will arrive, even on large islands.

Thus, a small island close to the mainland might easily have more species when it reaches equilibrium than a large island that lies far from the mainland.

Simberloff and Wilson tested the theory by tenting and fumigating tiny mangrove islands lying at various distances from a large island in the Florida Keys. They then measured how quickly the fumigated islands re-acquired insect populations, and how diverse their populations became. It is difficult to imagine such an experiment being permitted today!

All the islands eventually accumulated as many insect species as had been living on them before they were fumigated (though many of the new colonists were of different species). The equilibrium numbers of species that were reached on the islands of different sizes and at different distances from the mainland agreed with the predictions of the theory.

The theory of island biogeography has played a huge role both in ecology itself and in its new subfield of conservation biology. The theory shows us not only why small, distant islands tend to have few species, but also how and why species tend to be lost even in mainland areas if the mainland's natural habitat becomes fragmented. The theory predicts how such fragmentation of habitats—the equivalent of breaking them up into islands—will result in species loss, regardless of whether the fragmentation takes the form of separate islands in a body of water or of little bits of natural landscape that are surrounded and progressively squeezed by metastasizing city suburbs.

This ecological theory is simple and elegant, and it can predict consequences that are immediate and sometimes devastating when natural environments become fragmented. But it deals with the moment, and it makes no assumptions about evolutionary change.

At the time Darwin visited the Galápagos, he was more concerned with cataloguing its diversity than with the role of extinctions. Nonetheless, he encountered plentiful warning signs. Extinctions and near-extinctions, especially after the introduction of pigs and rats by visiting ships, had already begun to transform the islands. On his visit to Floreana Island, he heard from the archipelago's governor that years earlier the crew of a single frigate had, during a brief but devastating visit, almost emptied the island of its giant land tortoises. The governor feared that the tortoise population would not recover.

The few remaining tortoises were a fascinating example of the adaptation to each island's species mix that would help to lead Darwin to his deep understanding of how ecosystems evolve and diverge. These tortoises had extreme "saddleback" shells that let them stretch their heads upward. They could feed on the overhanging branches of the bushes that were plentiful on the island (and sparse on many other nearby islands). And, exactly as the governor had predicted, the island's unusual variety of tortoise did in fact go extinct two decades after Darwin's visit. Attempts are now being made to repopulate Floreana with tortoises from other islands that share at least some genes with those of the original population [17].

The MacArthur and Wilson model, superbly predictive as it is of such short-term changes, views islands as fragile vessels holding a temporary and ever-diminishing store of diversity that is only replenishable through immigration. As Darwin began to organize what he had learned from his voyage, his attention was focused on evolution's longer-term role as a generator of diversity through adaptive radiation on the different islands. The consequences of evolutionary change are not dealt with by MacArthur and Wilson.

It is clear that both adaptive radiation and the limits imposed by island biogeography are taking place in the world around us. To an observer of the current state of the Galápagos Islands, it would seem that the immediate negative forces of island extinction are overwhelming the achingly slow and tentative constructive processes of adaptive radiation. But perhaps the two processes are not as mismatched as we might think.

Ecology, Evolution, and the Coexistence of Species

Early ecological theory tended to treat species as immutable—an echo of the pre-Darwinian world. Mathematical ecologists constructed equations that were based on a manageable number of simple assumptions and were designed to predict the expected relative abundances of animal and plant species in the ecosystems that were being measured.

Some of these ecological principles seemed to be iron-clad. In 1934, the Russian ecologist and microbiologist G.F. Gause published elegant experiments in which he addressed a central ecological question: can two different species occupy exactly the same ecological niche indefinitely [18]? He started with pure cultures of two related, but clearly distinct, species of the tiny slipper-shaped microorganism *Paramecium*. He added equal numbers of these two species to flasks containing a suspension of yeast cells that the paramecia could eat. Some of the flasks had a fresh medium, while others contained a medium that had previously been used to grow one or the other of the species.

The two *Paramecium* species, *P. caudatum* and *P. aurelia*, normally feed on a variety of microorganisms, but in this experiment they were forced to compete for exactly the same resource. One of the two competing predators won the competition, but both could draw on different talents. The *caudatum* species grew more quickly and won on a fresh medium. The *aurelia* species won over *caudatum* when the previously used medium was employed, apparently because of its greater ability to access the resources that had been left behind by the previous users.

In the simple worlds that Gause constructed, in which each flask constituted a single ecological niche, it was clear that these two species could not share the same set of resources. But both species were able to win if they were presented with resources that suited their talents well.

Experiments carried out in the 1940s by A.C. Crombie went a step further [19]. Crombie put two species of flour beetle into a uniform world—a bottle filled with flour—as their only food source. He obtained the same results as Gause's experiments with *Paramecium*—one species always won. But when he added tiny bits of glass tubing to the flour, both beetle species could survive. He found that the larvae of the smaller species could creep into the little tubes and pupate there, avoiding the predatory larvae of the larger species.

Crombie also found that if he simply substituted grains of wheat for the much more uniform flour, he could maintain three beetle species for long periods. He was able to show that larvae of one of the three species were able to burrow into the wheat grains and pupate there safely. But he was unable to discover exactly how all three species were able to maintain a three-way equilibrium, because the beetles' age distributions and the changing state of the wheat grains added too many variables.

The three coexisting beetle species were inhabiting a slightly more complex world than the utterly uniform one filled with flour in which two species had battled it out for dominance. A more varied world of wheat grains, each with its own internal structure, clearly presented more ecological opportunities for the beetles.

These and other experiments led to theories that dealt with such complicated situations. For example, when food resources are complex, so that some fraction of them can be accessed by all the competing species and other fractions can only be accessed by one of the competitors, conditions can be found in which two or more of these species should be able to coexist indefinitely.

Haunting all these theoretical efforts was the possibility that these theories, like the carefully constructed laboratory environments, did not capture the complexity of the real world. And the properties of the species involved might themselves be changing with time. As we will see in Chapter 6, recent experiments by the Japanese researcher Nori Ichihashi show that a single exceedingly simple "organism," living on what is apparently a single resource, can quickly evolve into more than one "species" that divides up that supposedly single resource in unexpected ways.

In 1960, Robert MacArthur, the co-developer of the hugely successful theory of island biogeography, expressed concerns about the effect of evolution clearly—only to reject them as unimportant [20]. He began by dividing various ecological theories up into those that emphasized one or the other or both of two variables that are commonly used to predict species abundance.

The first variable is the number N_s of each species that are present in an ecosystem. The second is the rate at which a species increases or decreases in numbers over a specified time interval, expressed as r, the *intrinsic rate of natural increase* (or decrease) of the species.

Often, both variables play a role. Suppose you survey an ecosystem over time, and find that the species in it remain relatively constant in numbers, or perhaps that their populations fluctuate in a predictable way. This suggests that they are partitioning up the resources of the ecosystem so that each species has reached an equilibrium number N_s. The theories that predict the outcomes of the flour beetle competitions, which also assume constant rates of intrinsic natural increase, fall into this category.

But what about cases in which there is variation in the intrinsic rate of natural increase? MacArthur suggested that such changes can happen when species are presented with a new opportunity.

Confronted with the problem of explaining such cases, MacArthur simply threw up his hands. Opportunistic species, with properties that change as they are competing and otherwise interacting with each other, could produce any result you might imagine. He shrugged off such situations:

There is little ecological interest in the relative abundances of opportunistic species . . .

By dismissing or minimizing the importance of such opportunistic interactions, MacArthur was suggesting that natural selection has little effect on ecological interactions. But, as Darwin had realized, natural selection is ubiquitous in ecosystems, and is primarily the result of the fact that other species in the ecosystems are also evolving. This produces continually shifting interplays among species, opening up fresh opportunities for the individuals in the species that can take advantage of new conditions more effectively than others. And, in response, the other species are likely to evolve different new ways to compete. Viewed in this way, every one of the ten million or more species on our planet is opportunistic.

I do not want to stray into anthropomorphism, that distressing tendency to impute human attributes to other organisms. But the process of evolution does provide opportunities, often quite remarkable ones, for all species. Most species do not consciously take advantage of these opportunities, but the opportunities are there to be taken nonetheless.

Thus, if living species were to adopt a motto, they could do no better than choose the one adopted by the New York Tammany Hall politician George Washington Plunkitt (1842–1924). The corrupt political machine of Tammany Hall shaped the careers of many corrupt politicians, but few were as cheerfully corrupt as G.W. Plunkitt.

Plunkitt was the inventor of what he charmingly termed "honest graft," a promise that if you paid him enough to accomplish something worthwhile, he would be sure to see that it happened [21]. And his motto was:

I seen my opportunities, an' I took 'em!

As Darwin had realized, an entangled bank of interacting species is a cauldron of such shifting opportunities, waiting to be taken advantage of by the process of natural selection. And such an ever-shifting entanglement of opportunities does not lend itself to easy analysis by ecological theories, no matter how elegant. Entanglements' variables are numerous and change in unpredictable ways.

But does all this matter in the real world? Humans can do enormous damage to ecosystems over a span of decades. Surely evolution must be far too slow to have much of an effect on the fate of ecosystems during such a brief span of time. Have ecologists therefore been justified when they ignore the opportunistically creative process of evolution? I think not.

In the Introduction, I suggested that we are now poised to begin to unravel the rules that govern ecosystem interactions. We are learning how, and how quickly, evolution can drive genetic change. And we are finding that even when evolution is faced with the task of counteracting swift and remorseless ecological processes, it may not be as helpless as we fear.

But first, to set the scene, I will try to convey through a few stories why these explorations matter so much. The first tale in the next chapter is a personal one. I was introduced to the consequences of our thoughtless damage to the world's ecosystems at an early age. And I suspect that many readers of this book have had similar experiences, though details may vary.

The second tale in the chapter takes us on a journey back through deep time and to a far corner of the world, to illustrate how our species could have been responsible for widespread environmental damage, aided only by fire and a few primitive tools.

2

Lost Worlds

Où sont les neiges d'antan?

François Villon (1431–63?)

I was born in London in 1938—not the happiest of years in which to make an entrance. My father was a Canadian citizen, so my little family was able to emigrate to Canada in 1940. This enabled us to escape the Blitz, just as it was beginning.

Arriving in Canada with no resources, we lived a rather sketchy existence. We eventually found ourselves in Edmonton, Alberta. There my father, totally without qualifications but with an ability to get people to work together, became the head of procurement for the construction of the Alaska Highway Project. This was a vast joint effort by the United States and Canada to provide a route by which Alaska could be supplied with resources needed to prevent the Japanese from invading.

The highway crews on the project faced enormous challenges as, for the first time, they built a road across Arctic permafrost. The highway was mostly completed by the end of the war, but the invasion never materialized.

After the war, we moved to British Columbia, Canada's westernmost province, where my father found his true calling as a sales representative for manufacturers of men's clothing. In the late 1940s, I spent several of my childhood years on the rural shores of a northern branch of Burrard Inlet, the great harbor of Vancouver. The branch was known as the Indian Arm.

We settled for several years in the little hamlet of Dollarton. This cluster of houses and shacks on the western shore of the Indian Arm had been named after a Scottish-Canadian shipping and lumber entrepreneur, appropriately named Robert Dollar.

With the onset of the Sino-Japanese War and the looming wider war in the Pacific, Dollar's lumber enterprise on the Indian Arm ran into difficulties. The mill was formally shut down in 1942, five years before my family and I arrived. I knew nothing of this history, but I soon discovered, not far from our house, the foundation of a huge building surrounded by thick second-growth forest. Only a giant concrete cellar, open to the air and filled with rainwater, remained.

The cellar must have been what was left of one of the buildings of Dollar's lumber and shipping enterprise. Now it was a magical place, always filled with water from the plentiful rains. The water swarmed with tadpoles and frogs, an endless source of inspiration for a budding biologist. Indeed, frogs were everywhere in the world of my childhood. If the smallest container of water were left out overnight, it would be brimming with masses of frogs' eggs by the next morning.

The inlet itself was another enchanting place. Along its shore, multi-armed orange starfish close to 1 m across dominated the thriving intertidal zone. Digging among the rocks always yielded something new, such as giant green Nereid worms that could bite the finger of an incautious young investigator. Far out in the middle of the inlet, 1.5 m-long tyee-chinook salmon leaped clear of the water. They were on their way to the shallows of the small river at the head of the Indian Arm, which provided perfect spawning grounds.

I spent hours on the little wharf that had replaced Dollar's dismantled ship-loading complex. There I was able to peer down through the clear and sheltered water at the crowded marine ecosystem that thrived below.

Long before my arrival, Dollar and other lumbermen had cleared most of the big trees out of the area, but huge areas of thick secondary growth had quickly replaced the stripped land. This secondary forest stretched up to the remaining old-growth forest that still covered the mountains to the north. The forest's understory of prickly salmonberry bushes was dense and almost impenetrable, but it was possible to walk for long distances along the few dirt logging roads that remained.

The Tsleil-Waututh Nation of indigenous groups had depended, probably for thousands of years, on the plentiful fish, shellfish, and numerous crab species of the inlet. But even before I arrived, the fish and invertebrate populations were disappearing. An oil refinery had opened at the eastern end of Burrard Inlet in 1908, and by 1913, an oil slick had spread over much of the city's harbor. Such pollution has continued largely unchecked. The provincial and federal governments did belatedly recognize the rights of indigenous peoples to their original sources of food, but these rights have been interpreted in the courts as largely applying to resources that no longer exist.

Living as I was in the seemingly timeless world of childhood, I knew nothing of these events. I briefly visited Dollarton again twenty years later, and found that all of the secondary forest growth, along with the rural settlements of my childhood, had been swept away by sprawling suburbs that were already spreading up the sides of the mountains. Squatters' shanties along the shore, which had been colonized by deserters from visiting ships in the 1850s, had all gone. They had been replaced by substantial shorefront houses that—unlike the shanties—featured indoor plumbing.

The underwater world, too, had vanished. The intertidal zone was covered with gray ooze, and the starfish and myriad other species had vanished. The wharf that had dominated my childhood was still there, but the waters beneath it were now a murky, green, fishless wasteland.

At the time of my later visit, the tyee-chinook salmon had not been sighted in the inlet for years. On the eastern shore of the inlet, 3 km away, crouched a giant new pulp and paper mill that was chewing its way through the secondary forests that remained. The mill spewed untreated acidic effluent into the inlet.

By the end of my return visit to Dollarton, I felt fully the depression caused by the loss of seemingly essential parts of the world around me—a feeling that most of us have experienced during our lifetimes.

The Great Australian Die-Off

Everybody who, like me, has been around for a while has a similar story of living through environmental plundering and degradation. I was perhaps a little more sensitized than many. During my Dollarton childhood, I had become aware of the importance of both the above- and the underwater world. From hours spent hiking through the thick secondary forest and peering down at the dense schools of fish beneath the little wharf, I had gained some sense of the totality of an ecosystem. Most visitors at that time would have simply glanced at the thick forest growth and the surface of the inlet, then moved on.

What we are now doing to the planet fills the pages of newspapers and scientific journals. What is less known is how long we have been doing it. The farther back in time we go, the more fragmentary and hard to interpret the clues to these historical events become. But sometimes, clever science can use clues in the fossil and geological records to reveal a vivid glimpse of the effects of our species in the distant past. Our impacts during those early days seem often to have been far greater than should have been possible given our ancestors' small numbers and primitive technology.

Some of those early actions were devastating. Consider what the first humans did to Australia.

Genetic clues, and to a lesser extent fragmentary fossil evidence, have been used to trace the first great migration of *Homo sapiens* out of Africa. This migration started in the Horn of Africa about a hundred thousand years ago and ended in Australia and New Guinea. It progressed by many routes and by slow stages through a still-rich and verdant Middle East, crossing the Arabian deserts during periods when they were a bit less formidable than they are today [22].

The great delta of the Indus in western India must have provided a welcome oasis on the eastern side of these barriers. From there, adventurous bands were able to fan out across India. We know from genetic evidence that some of them traveled down the subtropical coast overlooked by the hills of the Western Ghats, and then around the southern tip of the subcontinent. They left their genetic traces behind in the tribal groups of the southern states of Kerala, Karnataka, and Tamil Nadu.

Some of these groups worked their way across the river system of the Ganges-Brahmaputra and down the vast peninsula of southeast Asia, fanning out into what is now the Indonesian archipelago. Aiding these migrations were periods of glaciation during which sea levels dropped and some of the islands were joined by temporary land bridges. A large fraction of their genetic legacy persists in the isolated tribes of the Andaman Island chain that extends south from Myanmar into the Indian Ocean [23].

These migrants left behind detailed representations of animals, humans, and imaginative human-animal hybrids in caves on the islands of Borneo and Sulawesi [24]. Some are as much as 44,000 years old. They are the artistic equal of any figurative cave art from France or Spain, and they are at least 7,000 years older.

On the long route from Africa, the migrants encountered some members of slightly divergent branches of the human family tree, who had long preceded them on their own earlier migrations out of Africa. Many of us still carry genes that we have inherited from encounters with various close relatives—the Neanderthals, whose ancestors had spread earlier to regions in western Europe and central Asia, the still-mysterious Denisovans who seem to have thrived in scattered groups from central to eastern Asia, and other peoples who are lost in time [25]. Some of those genes, notably an allele that adapts people to high altitudes, have been shown to aid survival [26].

The most adventurous of the modern human migrants traveled east along the Lesser Sunda Island chain, which extends from today's island of Bali east to Timor. From there they managed—no one knows how—to cross substantial stretches of water in order to reach New Guinea and later Australia.

Although tribes from New Guinea and its nearby islands would later play an important role in the peopling of the Pacific, the remote island continent of Australia was the furthest point that was reached in that first great migration.

We do not know when and by what route the very first human immigrants to Australia actually arrived. But recent genetic evidence suggests that some of the early settlers in New Guinea might have taken advantage of land bridges across the Torres Strait, which enabled some of them to make their way south to what is now northern Queensland. Many other paths to Australia remain possibilities [27].

Whatever their exact route, those first migrants to Australia spread through the island continent. They left few signs of that early spread, so perhaps they traveled swiftly. Indeed, the earliest signs of human occupation anywhere in Australia that can be unambiguously dated have been found in the continent's far southwest corner at what must have been the end of the great migration. The charcoal remains of cooking hearths in the dramatically named Devil's Lair Cave, near the coast of southwest Western Australia, are at least 48,000 years old [28].

Subsequent events have largely masked any impact that the participants in that great migration might have had on their environment during their journey across Asia. But lucky circumstances in the prehistory of southwest Australia have allowed scientists based in Australia, Europe, and the US, led by paleontologist Sander van der Kaars of Australia's Monash University, to gain new insights into the problem. They have pieced together clues that have tied animal extinction events in Australia directly to the arrival of the first humans [29].

These human arrivals were the seasoned survivors of thousands of brushes with death that their ancestors had experienced during the many generations of their long migration. The living world of Australia that greeted them was filled with a variety of large animals or *megafauna* [30]. The topmost predator was a "marsupial lion," *Thylacoleo carnifex*, an overgrown relative of the present-day wombats and the recently extinct Tasmanian tigers. (Unusually, many of Australia's carnivores seem to have evolved fairly recently from herbivore ancestors, who were also the ancestors of the highly adaptable wombats.)

This lion-like predator was as massive as the largest present-day placental lions [31]. It was equipped with huge shear-like premolars that could exert the largest bite force of any known predator, living or fossil. Unlike any other marsupial, it had retractable claws, suggesting it might have been able to climb trees. For those of you who have monster-filled nightmares, I recommend *Thylacoleo* as a worthy addition to your menagerie.

The marsupial lion had a variety of large prey from which to choose. The biggest was the hippopotamus-sized *Diprotodon*, a massive grazer which, like its leonine predator, was related to the wombats. Also potentially on *Thylacoleo*'s menu were sheep-sized, egg-laying spiny anteaters, giant marsupial kangaroos and wallabies, and the flightless Stirton's Thunder Bird, which weighed half a metric ton. More than 3 m in body height, the Thunder Bird may have been the tallest bird that has ever lived. Outsized reptiles included immense goannas and terrestrial crocodiles that could reach 7 m in length [32].

These Australian megafauna are now all extinct. Many of their smaller relatives still survive, but the larger animals have all vanished.

This is an unusual extinction pattern. On other continents, substantial numbers of species, both large predators and large prey animals, managed to persist during human colonization.

How involved might humans have been in these Australian extinctions? The disappearance of these large animals has been dated, using the fossil record, to roughly the time that the first human migrants arrived in Australia. But very precise times of extinction cannot be determined from the fossil record alone. This is because the last recorded appearance of an extinct animal might easily have pre-dated its actual extinction by thousands of years.

Similar but less sweeping waves of extinction in the Americas, and on large but isolated islands such as Madagascar and New Zealand, are also coincident with the arrival of humans. In the case of the flightless giant New Zealand moa, there is strong evidence for human involvement in its extinction. Analysis of the oldest traces of rats on New Zealand dates the arrival of the first rats (accompanying the first humans) to at the latest AD 1400. The last moas disappeared in about the year AD 1600 [33]. But the human arrivals to these regions were far better-equipped with hunting technology than the earliest humans of Australia, with their simple Stone Age technology. The Australians did have spears, and used spear-throwers that acted as powerful extensions of their throwing arms. But they lacked the far more accurate and versatile bows and arrows that could be used at a safer distance from fierce animals such as *Thylacoleo*. And at that early time, there were no hunting dogs to help them [34].

In many of the possible cases of human-caused animal extinctions, climate fluctuations cannot be ruled out as a contributing cause. Even in regions as warm as Australia, the climate has shifted repeatedly, from cooler and dryer to warmer and wetter and back, in synchrony with advances and retreats of polar glaciers. These climatic changes could have reinforced the effects of human hunting, or may even have been solely responsible for waves of extinction.

Sander van der Kaars' group needed new evidence that would help them to distinguish between climate and humans as causes of the megafaunal extinctions. They did have that fairly precise date, roughly 48,000 years ago, for the earliest arrival of humans in Australia's southwest (although, of course, the first humans might actually have arrived even earlier). They also knew, from studies of fossil plant communities, the approximate dates for those successive climatic changes between cooler and dryer and warmer and wetter that have characterized Australia's climate.

The last such change before the arrival of modern humans, from cool and dry to warm and wet, had been centered about 22,000 years earlier. This made it an unlikely cause of the extinctions.

There were already indications from previous studies that the Australian megafaunal extinctions had happened substantially later than that climate change. It was quite possible that the extinctions had peaked shortly after the documented arrival of humans. But those extinction estimates were based on statistical averaging of extremely noisy data sets, consisting of the often-uncertain dates of extinction of various animal species that had been obtained from fossil finds. In sum, the evidence did not rule out the possibility that humans rather than climate change were responsible, but it also did not provide a resounding vote in favor of humans.

Van der Kaars' group came up with an ingenious way to get better data. Around the world, the wind blows dust from the land out to sea. The dust falls to the water's surface and eventually drifts down to the sea bottom. There, it contributes a tiny amount to the slow buildup of bottom ooze, most of which consists of the skeletons of tiny planktonic organisms.

The world's seabeds are relatively quiet places, geologically. They tend not to be disturbed by erosion, or by those great upheavals that are a common feature of the tectonic plates of the continents. Thus, if one carefully lowers a drilling rig to the sea bottom and drills a core down into the soft deposits, the result can often be an undisturbed record of what has settled there, a record that can encompass hundreds of thousands or even millions of years.

Van der Kaars' group examined such a core from an undisturbed piece of seafloor lying 300 m deep and only 150 km out to sea from the Devil's Lair Cave. The core reached back 150,000 years, and showed that the bottom had been undisturbed during all that time. Embedded in the core, along with the remains of oceanic plankton, were millions of pollen grains, fungus spores, and bits of ash from fires, all carried there from the land by the wind.

Currently, this coast of southwestern Australia is heavily farmed, but of course these changes took place only after the arrival of European settlers. The area still harbors many native plants. Thousands of species of endemic wild-flowers still bloom in the spring, making the whole region a mecca for gardeners and amateur botanists.

On a recent visit during the Australian spring, after emerging from the treeless and mercilessly barren Nullarbor Plain to the east, I drove to the sea through great groves of jarrah, marri, karri, and tingle eucalypts and the blazing carpets of wildflowers that surrounded them. These forests and fields of flowers must still be bequeathing their pollen to the nearby seabeds.

More information was concealed in the core. Isotope measurements could date its layers precisely. The pollen grains, with their characteristically sculpted surfaces, could be identified down to the genus or even the species level. As van der Kaars' group traced the ways in which the abundances of flowering plant species changed over time, they could infer changes in rainfall and average temperature.

Most dramatically, changes in the abundances of fungus spores allowed them to make a surprisingly precise estimate of the total biomass of the terrestrial animals that were living in the adjacent land area. Many of the spores were from fungi that belong to the genus *Sporormiella*, which thrive on animal dung.

Spores of these fungi had already been used with some success to track fluctuations in overall mammal population size in various parts of the world. But in those earlier studies, the changes in spore numbers were measured in cores that had been drilled in lake bottoms and other land deposits. Such strata were far more likely to have been disturbed than the much calmer ocean bottom.

The number of *Sporormiella* spores in that ocean-bottom core stayed high and held quite constant through several big vegetation and climate changes. But they plunged almost threefold starting approximately 45,000 years ago, about 3,000 years after those earliest traces of human activity in Devil's Lair. *Sporormiella* numbers have stayed low ever since [35].

Do these data absolutely show human culpability in the extinction of Australian megafauna? Of course not. But they strongly support the hypothesis that the earliest human colonists of Australia may have had a huge and astonishingly rapid impact on the island continent's ecosystem. If humans were the cause, they were able to bring about these extinctions even though they were few in number and possessed only the most primitive Stone Age technology.

They had not, unlike later human migrants, been accompanied by dogs, cats, chickens, and other potentially damaging placental mammal companions. Indeed, Australia had been without any placental mammals through most of its history as a separate body of land.

A variety of species of rat did evolve in Australia, branching out from two groups of migrants from Asia about six million years ago [36]. Dingo dogs are much more recent arrivals, making their entry perhaps as little as 3,500 years ago [37]. Relatively recent human migrants to New Guinea had brought dogs with them, and some of the dogs had evolved there into the highland singing dogs. Some of those New Guinea dogs must have reached Australia, giving rise to the genetically very similar dingoes.

This emerging story is discomfiting on another level. The monotreme and marsupial megafauna of Australia seem to have been surprisingly helpless against the first human colonists. Marsupial vulnerability continues down to the present day. Extinction rates of the remaining native mammals have increased with the arrival of Europeans and the introduction of devastating placental predators, especially cats, black rats, and foxes, and of competing grazers such as sheep, cattle, and—of course—rabbits. The extinctions have reached 10 percent of the total native species, a rate ten times higher than the proportion of extinctions in North America since European colonization. Another 21 percent of the native Australian species are listed as threatened by the International Union for Conservation of Nature (IUCN), compared with an average of 3 percent for the rest of the world.

Puzzlingly, even the Australian species of rat, descended from rats that arrived before humans, are also going extinct at a rapid rate. In some cases, recently introduced cats appear to be primarily responsible [38].

At least since the separation of Australia from Antarctica eighty-eight million years ago, and probably earlier, Australia's evolutionary cauldrons have been bubbling away in isolation from the rest of the world. How did those ecosystems become so vulnerable?

Today we have incredibly sophisticated tools to monitor such endangered ecosystems. We can measure changes in water flow and weather patterns, through the use of satellites and in the distributions of hard-to-find animals and birds using remote camera stations, in the chemistry of the environment, and perhaps most importantly in genes.

We will soon be able to probe exactly why some ecosystems, such as the ancient, isolated ecosystems of Australia, are so unusually vulnerable. We may be able to discover why those Australian placental rat species have lost their ability to withstand the recent onslaught of placental invaders.

Which evolutionary entanglements show characteristics that are unique to endangered or fragile ecosystems? In particular, what components of these entanglements must we supplement, or restore, if we are to halt or reverse these losses? And how can we avoid making perhaps irreparable mistakes, throwing the endangered ecosystems even further out of balance, as we learn how to carry out these changes?

As we will see, there are poison pills concealed in such new knowledge. We must learn to read warning flags and messages about the limitations, hidden strengths, and hidden dangers facing the living ecosystems that we and our ancestors seem to have been able to damage so easily. If we succeed in doing this, we may avoid losing an entire planet.

3

How Ecosystems Survive Change

The origin of the entire [Pomotu] group is generally ascribed to the coral insect.

According to some naturalists, this wonderful little creature, commencing its erections at the bottom of the sea, after the lapse of centuries, carries them up to the surface, where its labours cease [. . .] Here and there, all over this archipelago, numberless naked, detached coral formations are seen, just emerging, as it were from the ocean. These would appear to be islands in the very process of creation—at any rate, one involuntarily concludes so, on beholding them.

Herman Melville, *Omoo: Adventures in the South Seas*, 1847

With national budgets now habitually measured in multiple trillions of dollars, and with telescopes able to reveal events that took place billions of years ago and billions of light years away, we are gradually becoming used to the astronomical numbers that are needed to describe the universe around us. Even so, it is still difficult for us to come to grips with spans of time greater than the few thousand years of written human history. What was the world really *like* a million years ago—and, far more unimaginably, what will it be like a million years from now? And what evolutionary processes were—and will be—involved? In this chapter we will explore the amazing changes that have happened to one type of ecosystem, that of the world's tropical reefs. And this will set the stage for us to probe how these changes happened.

We now know, as our ancestors did not, that evolutionary and ecological change has been taking place over billions of years, not mere millions. But the path to this realization was a tortured one.

The world of most of Darwin's contemporaries was far more circumscribed than ours. It was, for the majority of Europeans, confined within the highly suspect dating of the Old Testament by Archbishop James Ussher (1581–1656). He had put the time of Creation at an amazingly precise October 23, 4004 BC (using the old Julian calendar). This was a little earlier, and suspiciously much more exact, than the Jewish traditional date of 3761 BC.

Of course, not everyone believed in that circumscribed world. Darwin's friend and colleague Charles Lyell and other geologists were already surmising that the Earth must have a very long history, certainly going back many million years.

Excavations for canals and railroads were uncovering the Earth's subterranean secrets as never before. These discoveries were not without cost. In 1853, Hugh Strickland, a lecturer in geology at Oxford, was killed by a passing train as he examined the strata that had been made visible on the sides of a new railway cutting.

In England and Scotland, these excavations revealed the results of slow continuous processes of erosion and deposition. In 1788, Scottish geologist James Hutton had famously encapsulated this *uniformitarian* view of geological history: "No vestige of a beginning, no prospect of an end!"

Darwin realized that enormous spans of time were necessary for life to have evolved through what must have been innumerable slight changes that separated its simple beginnings and its present complexity. But hard physical evidence from the rate of our planet's cooling appeared to show that the Earth was young, which posed a great challenge to his theory.

This hard evidence, based on physicist Lord Kelvin's calculations, was based on precise numbers. But the calculations had not taken heat from radioactive decay into account [39]. The discovery of natural sources of radioactivity changed all that. Since Kelvin's time, many dating methods have become available to scientists. Some of these methods can date rocks that are billions of years old, by using the decay products of long-lived radioactive isotopes in old rocks and measuring how long it has taken for one isotope to decay to another since the time when the rocks formed. Shorter-lived isotopes in volcanic ash and lava can be used to date past volcanic eruptions with great precision.

The number and type of tracks in crystals that are left by radioactive decay often provide essential clues. Vast events such the periodic switching of the Earth's magnetic poles, resulting in reversals in the direction of the residual magnetic field in successive layers of rock, can often yield dates in places where other methods are not available. Even recent events can be dated, by using the amount of carbon-14 in old organic materials or the patterns of wide and narrow growth rings in buried tree trunks. The old cave paintings that were recently discovered in Borneo and Sulawesi were dated by measuring radioactive decay products in the succession of layers of transparent calcite deposits that had built up over the pigments of the paintings.

The field of dating continues to grow in sophistication and accuracy. For example, the moment in time when the dinosaurs died as a result of the great Cretaceous-Tertiary asteroid impact was recently revised from sixty-five million years ago to sixty-six million. And, spectacularly, a deposit of fish fossils from that event has been found in North Dakota. The fishes' gills were filled with glass tektites created by the meteorite impact 4,000 km to the south. The fish appear to have died at the instant when the impact-caused tsunami, filled with tektites and still gigantic, arrived at their location. Their bone growth patterns have narrowed the asteroid collision to a northern hemisphere spring [40]! Such precision is not quite in Archbishop Ussher's league, but it is close—and it is based on much more convincing data.

When two or more different methods are available to date a particular sample of rock or a fossil, they almost always agree. Such precise dating, along with our growing knowledge of the fossil record, allows us to locate specific events in the past with growing confidence. And paleontologists have long moved beyond simply digging up (or stealing) dinosaur skeletons and ignoring their context, as the larcenous dinosaur hunters of the nineteenth century had done. The planet is being scoured as never before to find collections of fossils that give remarkably complete pictures of entire ecosystems of the past. Tracing out such histories can show us how ecosystems have changed, and how resilient they can be.

Sometimes, these ecosystems leave exquisitely detailed records. Because of their stone-like composition, corals reefs leave an abundant fossil record that can be used to trace their eventful and surprisingly resourceful histories.

How Tropical Reefs Have Adapted to Change

In 1960, soon after I had learned to scuba dive in the frigid waters of British Columbia, I dove on a tropical coral reef for the first time. It was not just any reef, but one of the world's greatest barrier reefs.

This reef runs the length of the western margin of the deep "Tongue of the Ocean," a great underwater canyon that separates the Bahamas' largest island, Andros, from the rest of the Bahamian archipelago.

On one of our first dives together, our group of divers swam down the seaward slope of the reef, leaving behind a forest of staghorn coral that swarmed with schools of French grunts and other brilliantly colored fish. We traveled across a gradually deepening sandy bottom, dotted with immense cathedral-like columns of coral so tall that they nearly broke the surface. Then, at a depth of 30 m, the bottom plunged away in a heart-stopping cliff.

We peered straight down a precipice that stayed vertical until everything faded from our sight into a vast expanse of deep blue. Sonar tracings have shown that this cliff falls straight down for 1,000 m and then slowly transitions into a steep slope until it reaches the bottom of the Tongue of the Ocean at 2,000 m.

We ventured down the wall to a depth of 60 m. From today's perspective, almost two-thirds of a century later, this was a really stupid thing to do. We had primitive dive gear, were beginning to feel woozy from nitrogen narcosis, and had no dive computers to calculate decompression times. I have only dived half as deep since.

What we saw seemed to be worth the risk. The water was so clear that we could still see the waves far above, forming tiny wrinkles on the small circle of the surface that was still visible directly overhead.

Half of our world was made up of wall, and the other half was open water stretching off into an unbelievably blue infinity. Every square centimeter of the wall was covered with life, ranging from tiny tunicates to long whip corals. Large pelagic fish could sometimes be glimpsed swimming far out in the blue.

I was hooked. Ever since then, I have seized any opportunity to dive the world's reefs. But it is heartbreaking to see how, during successive visits over the years, reefs have become progressively more damaged from sewage and agricultural runoff, warming water, dynamite fishing, and the clouds of sediment that are raised by fleets of fishing vessels as they drag giant trawls across nearby sea bottoms.

Today's reefs are complex, and they can be huge. The Andros barrier reef, down the vertical cliff-edge of which I had so injudiciously plunged so many years ago, is only the world's third largest. The Great Barrier Reef, the most gigantic and one of the most endangered, covers a third of a million square kilometers of shallow ocean bottom that spans much of Australia's east coast.

Reefs are so obviously made up of a complex entangled web of species that an observer can be forgiven for assuming that they must be very old. But in terms of geologic time, today's reefs have evolved surprisingly recently. Most reefs of the past were built by organisms other than corals.

Herman Melville, in the excerpt that begins this chapter, got the organisms responsible for today's massive constructs hilariously wrong. But he got some parts of reef construction right. Modern reefs are built up, as Darwin had known, not by insects but by successive layers of colonial coral polyps. These are tiny barrel-shaped organisms, each of which possesses a tuft of stinging tentacles to catch prey.

The polyps build protective and supporting pockets of limestone around themselves. Different species can knit their versions of these tiny limestone structures into a variety of shapes, from whip-like to antler-like to huge plates, all the way up to giant boulder-like colonies covered with convolutions that dominate the sea bottom like the brains of immense aliens. Other corals are "soft," and able to dispense with skeletons. They grow into flexible feather-like structures that come in a dazzling array of colors.

All the corals that live in the sunlit regions near the ocean's surface can catch prey, but most of their energy comes from tiny one-celled photosynthetic algae (zooxanthellae) that live inside the cells of the polyps themselves. Surprisingly, many coral species are now being found that live in much deeper waters, beyond the reach of light. These corals have no symbiotic algae. Their deep-water reefs, too, can be diverse. But, because their corals do not gain energy directly from the sun, they are not as exuberant or as fast-growing as those nearer the surface.

Darwin distinguished several types of coral reef. There are fringing reefs such as those that surround mountainous islands in the middle of the ocean. There are barrier reefs that can protect large masses of land, such as the great reef I dove on that shelters Andros Island from hurricane damage. These barriers can build up on shallow continental shelves. And then there are the most challenging to understand of all: atoll reefs.

A new volcanic island in the tropics can form, even in deep water, through repeated eruptions of magma from the hot mantle below. These eruptions pierce the relatively thin layer of crust that makes up most ocean floors. When such a volcano, growing beneath tropical seas, breaks the surface for the first time, its summit is suddenly surrounded by warm water and bathed in abundant sunshine.

Because the tiny offspring of coral polyps can swim and drift for long distances, even a new island of this type will quickly acquire a fringing reef of corals. The result is an ever-growing and ever-more-complex world of nooks, crevices, shelters, and mazes that soon attracts an attendant entourage of other species drawn from the entire tree of life.

Darwin was the first to suggest that such a newly formed volcanic oceanic island may easily sink again beneath the waves as it erodes away. We now realize that the huge mountain itself, once it stops erupting, will also sink from its own weight. As the mountain sinks, the corals that surround it will continue to grow upward, striving to stay in the zone of bright sunlight. Eventually, the island itself disappears, but the reef continues to grow up toward the light.

The most vigorous growth of these reefs takes place on their outer margins, where oceanic currents replenish plankton food supplies and carry away waste. The lagoon that forms in the center fills up with broken coral from storms, along with sand that has been defecated by parrotfish and the other reef animals that munch on the corals themselves. The eventual result is a ring-like atoll, surrounding a shallow lagoon, with no sign of the island that used to form its middle. The outer margin of the atoll drop away steeply.

The correctness of Darwin's theory was first tested during the 1897–8 expedition to Fiunafuti Atoll in the Ellice Islands. The boat carried equipment for drilling into the reef. The deepest of its boreholes reached less than 100 m, but the drill penetrated through layer after layer of coral just as Darwin had predicted.

The full ability of corals to transform the Earth's surface over millions of years was revealed as a result of horrific events in the mid-twentieth century. After the Second World War, Bikini and Enewetak (formerly Eniwetok) atolls, along with several other coral atolls in the Pacific, were used to test atomic weapons.

Just before the US' largest detonations, the Geological Service drilled boreholes in the reefs to conduct geological surveys. The boreholes at Bikini were drilled down through 750 m of coral without reaching bedrock. The deepest borehole at Enewetak reached 1.4 km before it finally encountered volcanic basalt far below [41].

The fossil corals at the bottoms of these boreholes had lived as long as forty million years ago. They had been badly deformed by pressure, but it was still possible to see that some of them were close relatives of present-day corals. Just as Darwin had postulated, the volcanoes that lay far below these atolls had subsided slowly beneath the surface, allowing uncounted generations of corals to reach for the sunlight by growing on the piled-up calcium carbonate skeletons of their ancestors.

Forty million years are a tiny slice of time in the history of tropical reefs. If you could have snorkeled through the Eocene reefs of forty million years ago, they would probably have looked very much like the reefs of today. Fish, undoubtedly brightly colored to attract mates, would have darted among branches and mounds of coral that looked like those we delight in swimming through today. But those early reefs had just emerged from a period of evolutionary turmoil. And that period in the long and complex history of reefs had been preceded by other vast upheavals.

The Reefs of Today

Today's reefs are superbly adapted to a world without extreme heat in the oceans and without excessive amounts of carbon dioxide in the atmosphere. But, as we are seeing daily from news reports, they are still teetering on a knife-edge. Two critical environmental parameters are changing more rapidly than they have done for millions of years.

Shallow-water corals compete strongly for light, and they must be able to grow upward to do so. To build their skeletons, most of the corals that make up the reefs need a source of the bicarbonate ion, HCO_3^-.

This ion results from the dissociation in water of carbonic acid, H_2CO_3. Dissolved CO_2 reacts with water to form carbonic acid, which immediately dissociates into bicarbonate and hydrogen ions.

The concentration of this acid, and the overall acidity of the water, are therefore strongly dependent on the amount of CO_2 that is dissolved in it. The more CO_2 in the water, the more acidic the seawater becomes.

Bicarbonate is a two-edged sword. The corals can combine bicarbonate ions with calcium ions, and fit the resulting calcium carbonate molecules together into a strong and flexible lattice of the mineral aragonite. This is easy for the corals to do, provided that the seawater is not too acidic. But when acidity is too high, the production of these strong and insoluble molecular chains becomes much more difficult and energetically demanding.

It was corals with skeletons made of aragonite that tore through the oak hull of the *Endeavour*, Captain Cook's ship, when it ran up onto Australia's remote northern Great Barrier Reef in 1770. The water that rushed through the jagged hole in the *Endeavour*'s bottom would have immediately sunk the ship, as similar accidents had doomed innumerable other ships, were it not for the brilliant leadership of the captain and the superhuman efforts of the crew. It took them six weeks to plug the hole in the hull with a sail, maneuver the ship free of the reef, beach it, then careen and repair it.

Coral reefs have terrified sailors for centuries, with good reason. Even ships made of tough and resilient oak could be pierced by outcroppings of even tougher and more resilient aragonite.

Luckily, the hull of a little Solomon Islands dive boat that my wife and I were on in 2006 was made of steel rather than oak. On the night of August 11, the steersman, who had been surprised and confused by a sudden squall, lost his bearings and ran us straight up onto a fringing reef.

When the sun rose the next morning, everything was eerily still. The ship, though undamaged, was truly stuck. It had, unluckily, run aground during the highest tide of the month. To make things even more embarrassing, kids from a nearby village were swimming around us in the sparkling water, laughing with delight at our obvious discomfiture.

Tugboats eventually arrived from the capital, Honiara, but several attempts to pull us free failed. Finally, faced with our own deadlines, we passengers had to leave and take a rescue boat back to Honiara.

The dive boat was eventually refloated through the efforts of three tug-boats—a testament to the gripping power of those jagged aragonite skeletons. The boat still plies the rich, teeming waters of the Solomon archipelago's remote islands.

Small comfort to the ship's crew, perhaps, but such an incident would not have occurred if they had been sailing on oceans in the far distant past.

Fighting for Light in the Past—Evolution in its Most Transformative Mode

Reefs of some kind extend far back into Earth's history. We can get a vivid glimpse of what those early times might have been like by visiting Shark Bay on the coast of western Australia.

In the warm, shallow, and highly saline southern part of the bay, mossy-looking lumps of what appear to be growth-encrusted boulders dot the clear water. They are not rocks. They are stromatolites, which look like rocks but are actually built up by successive layers of bacterial colonies. Many of these bacteria are able to photosynthesize, and use some of that energy to surround themselves with sticky polysaccharides. As they multiply and grow toward the light, they trap and bind particles of sediment in their polysaccharide coatings. These layers spread to cover their predecessors just as corals do.

Those smothered predecessors and their entrapped detritus form a support for the living layer. As each of the layers expires, it lifts the top active layer a tiny bit toward the light. Some of the bacteria are able to form limestone deposits.

Their layers tend to be in the form, not of the strong and flexible aragonite, but rather of calcite. This form of calcium carbonate is much weaker but less soluble than aragonite. And many marine animals can make calcite readily, even in acid seas.

Layered stromatolites thus presage, in a very limited way, the sophisticated coral structures of today. And even the earliest stromatolite formations played the same centrally important role as today's dead coral foundations—they lifted the living layers that had smothered them toward the light.

Many of these early stromatolite-like structures, found in some of the Earth's oldest rocks, may be non-biological in origin. But not all. There is evidence that living organisms may have been responsible for the traces of much-metamorphosed carbonates, which had almost certainly included calcium carbonate, that have been detected in 3.7 billion-year-old Greenland rocks [42]. If so, this would show that microbial communities were evolving on our planet a mere 800 million years after our planet's initial formation. And calcium carbonate buildup might have been lifting these communities toward the light billions of years before the appearance of corals.

The bacteria that are building today's Shark Bay stromatolites are, of course, only remote relatives of those ancient bacterial lineages. But their jobs are similar. The photosynthetic ones must be able to take advantage of the strong light in the shallow water, so that they can use the light to make energy-rich molecules. Other bacteria that are living in these communities can utilize surplus food molecules to help build and support the scaffolding of the stromatolites, and perhaps to defend against invading cells and viruses. Bacteria can form complex, multispecies mats on a variety of surfaces for a variety of reasons, but mats able to perform this whole range of functions are rare [43].

The details of these interactions, and how independently evolving species can interact in ways that lead to long-term ecological stability, remain largely unknown even in present-day stromatolite communities [44]. What is clear is that stromatolites of today can only thrive in a few extreme marine or estuarine environments. Those once-ubiquitous communities have now largely been replaced by other reef-builders.

The early stromatolites must have been even more limited in their capabilities. Their types of photosynthesis could not generate free oxygen. That would come later, with the unique and game-changing evolution of the ability to photosynthesize using only water, CO_2, and light—and to make free oxygen in the process. And the appearance of that ability, which we will examine in detail shortly, shows evolution at its most transformative, utilizing mechanisms of genetic recombination and rearrangement that have only recently been discovered. This seminal event opened the way for the evolution of reefs made up of more complex organisms.

This gigantic evolutionary leap was not something that happened within a single lineage of organisms. Ecosystems, with their complex sets of interactions among species, played an essential role.

Oxygenic photosynthetic ability appeared in ancestors of the Cyanobacteria, a group of photosynthetic bacteria that play a central but largely invisible role in the Earth's ecosystems. You have probably encountered some of them in the form of the green patches that often grow on earthenware plant pots.

Free oxygen had been present in only the tiniest traces for at least the first billion and a half years of the evolution of life. The descendants of the microbes that lived during that low-oxygen time, now mostly confined to oxygen-free environments such as swamps filled with decaying plants, are often actually poisoned by the merest trace of oxygen.

The evolution of oxygenic photosynthesis was probably the biggest thing to happen to life on Earth since the appearance of the first living cells. Slowly at first, and then more quickly, this new type of photosynthesis produced free oxygen that transformed the Earth's atmosphere and oceans and made multicellular organisms like ourselves possible. Today, most terrestrial and aquatic ecosystems depend on this process.

Oxygen production through photosynthesis is still solely carried out by the descendants of those early cyanobacteria, the first oxygenic photosynthesizers. Some of the descendants came to live in the progenitors of the green plants and of the zooxanthellae that live in the coral polyps. They have evolved a new identity as chloroplasts, symbiotes that live inside the cells of their green plant hosts.

The evolution of oxygenic photosynthesis makes possible every breath that we take and every mouthful of food that we eat. In Chapter 9, we will be ready to follow in detail how this transformative ability evolved. We will glimpse the incredible capacity for change that is possessed by the world's evolutionary cauldrons, and that led to such dramatic events. But for now, let us see how this advance shaped the evolution of the world's reefs.

The oxygen-driven environmental change was slow at first because free oxygen levels were kept low as oxygen began to combine with oxidizable minerals that were everywhere in the pre-oxygen world. But, over billions of years, stromatolite communities did manage to become more complicated. As oxygen molecules slowly increased in concentration, and as multicellular organisms evolved and themselves became more complex, these more elaborate organisms added to the abilities of those early reef communities.

The first of these more reef-like ecosystems may have begun to appear almost half a billion years before the beginning of the Cambrian [45], which itself began more than half a billion years ago. Sponges, which can have either glass- or calcite-based skeletons, began to play a larger role in reef structure. Fossils of one vast but now-extinct branch of the sponges, the Archaeocyathids, had calcite skeletons. Their fossils are often found associated with a rich variety of other fossilized organisms, and these ancient sponges may have provided support and protection for them. A time-traveling snorkeler visiting such communities, especially one who is familiar with sponge-rich Caribbean reefs, would probably recognize those ancient communities as becoming distinctly reef-like.

The first of these sponge-based, more reef-like ecosystems began to appear perhaps half a billion years before the beginning of the Cambrian [45]. This geological period was named after the Roman name for Wales, where these early fossils are common.

The beginning of the Cambrian is marked by an amazing apparent explosion of fossil marine organisms, many of which are clearly related to present-day multicellular animal lineages. Deep branches of the DNA family tree show that these lineages must actually have had a longer history, extending much further back than the start of the Cambrian. The DNA evidence shows that present-day animal lineages could not all have suddenly begun to diverge at the beginning of the Cambrian, but must have started to separate from each other 200 million or more years before. During those early times they were probably still small and soft-bodied and so that they left few traces in the fossil record. At the beginning of the Cambrian, they became larger and began to leave behind remains that were already mineralized (such as shells) or that could be mineralized (like bones and exoskeletons and impressions of entire bodies).

Fossil traces supporting these long divergence times are being discovered. Some tiny soft-bodied Precambrian animals of the newly discovered phylum Vetulicolia seem to be precursors of the later animal subkingdom that includes ourselves [46]. Other mysterious late Precambrian fossils called *Kimberella*, which seemed at first not to have had any later relatives, may actually be ancestors of the great phylum of the soft molluscs [47]. As paleontologist Simon Conway Morris memorably put it, the Cambrian explosion appears to have had a long fuse [48]!

There is evidence that a brief interval, during which atmospheric oxygen reached moderate levels, played a role during the most rapidly diversifying period of this apparent Cambrian explosion [49]. More oxygen would have made it easier for large—and hence fossilizable—multicellular organisms to evolve. But soon oxygen levels fell again, and most of the early stromatolite-like and sponge-based reefs disappeared—or at least they have not been preserved.

It seems likely that around the time of the Cambrian, the low oxygen levels that had kept organisms small and simple for billions of years were exceeded only slowly and fitfully. Nonetheless, the fossil record shows that reef ecosystems were becoming able to go through one transformative change after another.

Attempts have been made to imagine what reefs through the ages might have looked like. In 2002, Claudia C. Johnson of Indiana University presented a particularly sweeping view [50]. Her paper summarized the whole span of reef evolution. An illustration from her paper suggested how the appearance of reefs has changed, from the early stromatolite reefs that were common before the Cambrian down to the present day. The two parts of the picture take us through this entire history.

The figures in Plate 3 show the appearance and disappearance of the stromatolites and sponges, the rugose, tabulate, and early scleractinian corals, the massive rudist bivalves, and the many other reef-building organisms and collections of organisms that in succession provided the supporting skeletons of oceanic reefs during the last billion or more years. A sampling of some of the common smaller inhabitants of the reefs is also shown in Plate 3.

Our oceans have hosted living reefs of a wide variety of kinds, and early reef-building activity started long before the multiple origins of bacterial multicellular life more than two billion years ago [51]. And the evolutionary forces acting on reefs have often been surprisingly different from, and surprisingly out of step with, those that are acting on other ecosystems.

Extinction events that reefs could survive with ease sometimes brought other ecosystems to their figurative knees, and sometimes the reverse was true. Early coral and sponge reefs did well during two great mass extinctions, the Permo-Triassic and the Triassic-Jurassic, that bracketed the first period of the Mesozoic Era, the Triassic. Later in the Mesozoic, dinosaurs and early mammals thrived on land during a period of magnesium-rich oceans, when the high magnesium levels made it difficult for coral reef-builders to lay down aragonite [52].

Small environmental changes could sometimes bring about sweeping changes in reef species composition. These changes must have altered the reefs' entire evolutionary cauldrons.

Throughout this whole history, the evolutionary resources provided by the evolutionary cauldrons of the time were essential to reef survival. We owe our very existence to the fact that most ecosystems on our planet have been as varied and resourceful as these reef communities. But we are only beginning to glimpse why.

Are Today's Reefs Especially Remarkable? The "Moving Goalposts" Effect

I strongly suspect that today's dazzling reefs are more diverse than the reefs of the past. Richer evolutionary cauldrons should allow their component species to evolve in more new directions. But many ecologists would disagree. Those ecologists assume that there must be absolute limits to diversity.

Such absolute limits do indeed emerge from the use of ecological models that ignore the irritatingly unpredictable effects of evolution on what would otherwise be well-behaved, mathematically tractable ecosystems. But some data on ecological diversity have clearly suggested that these models are wrong. And, as we will see later, more data are beginning to suggest *why* they are wrong.

We saw that reef communities that depend on photosynthesis to power much of their growth, and their ability to support a diversity of interacting species, began as primitive reef-like structures that lasted through much of the early history of life up to the Cambrian. During that time, the resourcefulness with which organisms could exploit photosynthesis also increased enormously. Because of such evolutionary changes, even the earliest reefs gradually acquired a greater capacity to support ecological diversity.

The later history of reefs is a tale of gigantic changes in the mix of species that made them up. Most of the environmental pressures that drove these changes are lost in the past, but sometimes we can get glimpses of them. The shift to high global temperatures, which was driven by high CO_2 levels at the start of the Cretaceous and which led to a spike in ocean acidification, enabled the dominance of the calcite-forming rudist bivalves. These relatives of the oysters and clams grew massive enough to support entire reef structures. We don't know many of the details of those rudist reef communities, but they are likely to have been very different from the coral-based reefs of the present day.

Alas, the process of fossilization is mercilessly selective. Small, rare, and soft-bodied organisms tend to be underrepresented in the fossil record. Short of using a time machine, we cannot measure the true variety and complexity of those ancient ecosystems.

It is possible that even the oldest of those reefs were stunningly complex, like the reefs of today. But I suspect that they did not support as much diversity. Plate 5 shows a photograph that I took in 2019 of one of the many diverse reefs in the Indo-Pacific Coral Triangle.

The Coral Triangle spans both coasts of peninsular southeast Asia, along with the Indonesian archipelago, Papua New Guinea, the Philippines, and the Solomon Islands. It includes the world's most diverse coral reefs. The reef in the photograph lies just off the island of Misool, near the western tip of New Guinea.

My picture shows a tiny sample of the 4,000 species of coral and a few (I count twenty-nine in the picture) of the more than 2,000 species of fish that live in the reefs of the Triangle. You can also see soft and hard corals, hydrozoans, sponges, and a variety of those couch-potato relatives of ours, the sessile, filter-feeding tunicates.

The picture is an overview. The closer you get to such a reef, the more different creatures you will discover. And, as we will see later, this is only the visible world. DNA sequence analysis is revealing the millions of bacterial and virus species that also inhabit such complex ecosystems.

Were I, on the basis of such limited observations, cavalierly to condemn early reefs to a lower level of diversity than those of today, I would be guilty of a narrow chauvinism that exalts the wonders of the present day at the expense of the past. But a growing body of data suggests that I would be right.

Measurements of how (or even whether) the diversity of reefs has increased over geological time are fraught with difficulties. Such measurements depend on analyses of the fossil record—and that record, as we have noted, is often woefully incomplete. At the extreme, all traces of entire ecosystems in the past may be missing from the fossil record—and the likelihood that entire ecosystems might be overlooked increases the further back we go.

The ecosystems of the distant past must also have swarmed with untold numbers of small, soft-bodied, poorly fossilizing species. Even in fossil beds with well-preserved ancient reefs, such species could easily be missed. And of course we know nothing about those reefs' tiniest organisms, their bacteria and viruses. Those early ecosystems, even though they may appear simpler than those of the present day, could quite easily have been as diverse, and perhaps even more diverse, than today's ecosystems.

But some groups of organisms have left plentiful, detailed fossils. Studies of these groups do strongly suggest there has been an overall increase in ecosystem diversity—at least of those groups—over time. Perhaps the clearest signal of this trend has been obtained by my colleague Kaustuv Roy and his collaborators, who looked in great detail at the fossil record of marine bivalves [53].

When we hear the word "bivalves," we think of clams and oysters, but this taxonomic group also encompasses a huge diversity that includes burrowing shipworms and those rather boring but numerous rudists that interlocked with each other to make reefs during the latter part of the Mesozoic.

The shells of different species of bivalves differ in many details. But, because bivalve shells are in effect already made of rock, they can easily become fossils that preserve all of those details intact. Together, the living and fossil bivalves make up the clearest and most complete set of branching evolutionary lineages that the fossil record has so far bequeathed to us.

Fossilized shells of early bivalves first appeared in the early Cambrian, more than half a billion years ago. The number of genera (groups of species) of bivalves has increased, sometimes dramatically, ever since. This increase was almost unstoppable. Even the worst of the great mass extinctions of the past reduced the number of bivalve genera by less than 50 percent. The shelled animals soon recovered and diversified again after such disasters (Figure 3.1).

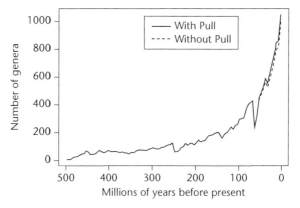

Figure 3.1 Increase in the number of bivalve genera from the Cambrian to the present (redrawn from part A of Fig. 1 of [49]). The solid line shows the uncorrected rate of increase, which might be biased by the "pull of the recent" (see text). The dotted line shows the increase after this possible biasing factor has been removed.

However, at least some of this exploding diversification might be the result of a statistical artifact, the "pull of the recent." This possible source of bias was first named and discussed by David Raup in 1979 [54].

Apparent increases in diversity may simply be because the member species of living ecosystems can be censused in their entirety, giving a picture of these organisms' true current diversity, while the less complete fossilized ecosystems cannot. This artifactual "pull" provides a possible explanation for the diversity increase. In effect, when above I favorably compared the diversity of living reefs with what appears to be the lower diversity of fossil reefs, I could have been misled by the pull of the recent.

In the case of the bivalves, this statistical bias can be corrected for to some degree. Roy and his colleagues redid their analysis, this time counting only bivalve genera that are known to have gone extinct and leaving out all the genera that include still-living species and that might be more diverse because the diversity of still-living species is included. Because bivalves have left a trail of numerous, superb, and unmistakable fossils, even from the oldest strata, this reanalysis could be done easily without losing much data. After the correction, the statistical effect of the "pull of the recent" turned out to be small. Removal of the pull (dotted line in the figure) hardly changed the dramatic increase in the numbers of genera during the last half billion years.

Thus, the dizzying recent climb in diversity, which is also apparent in bivalve data collected from specific parts of the world, seems to be real. But the feature of these data that I find most intriguing is the steepness of the recent climb in diversity.

Bivalves, of course, might be a special case. Increases in diversity are observed in many other groups, but not to such an extent as in the bivalves. And there could easily be groups of poorly fossilized organisms that have become less diverse with time. There could indeed be an upper limit to the diversity that our planet's ecosystems can support. It has been suggested that ecosystems have repeatedly bumped into that limit during our planet's history. But Roy and his colleagues do have strong evidence indicating that—at least for bivalves—marine ecosystems may even today be a long way from reaching such a limit. Why?

My guess is that the limits to diversity are continually shifting over time. It is possible that diversity did reach limits in the past, but that the limits shifted upward as entangled reefs—or forested banks, rainforests, or planktonic communities—grew in complexity.

Why should the limits increase? When new ecological niches open up, so do the opportunities for newly evolved or recently arrived species to take advantage of them. As soon as new mechanisms for defense, offense, or cooperative behavior evolve, ecological niches that can be utilized by new species will appear de novo or will arise from subdivisions of older niches.

As species increase in number, the possible interactions that can take place between them should also increase in numbers. This opens up opportunities for even more between-species interactions. In fact, the number of such possible interactions between species should increase much more quickly than the number of species.

Tropical reefs, as we have seen, are amazingly adaptable, able to respond to immense changes in their environment. Why? And do we also find such diversity and resilience in less obviously diverse ecosystems? To begin our exploration of these questions, we must look at the contents of the world's evolutionary cauldrons.

4

The Genetic Contents of the Evolutionary Cauldron

Eye of newt, and toe of frog,
Wool of bat, and tongue of dog,
Adder's fork, and blind-worm's sting,
Lizard's leg, and owlet's wing,—
For a charm of powerful trouble,
Like a hell-broth boil and bubble.

The witches again (*Macbeth*)

Shakespeare drew most of the hair-raising contents of his witches' cauldron from obscure names or code names for plants. Thus, the brew is a little more benign than it first appears. Evolutionary cauldrons' contents are generally even more innocuous, and often life-giving. But those contents, like the contents of the witches' cauldron, are extremely complicated and not without their dangers. Here, I want to trace their story.

Darwin knew from his own observations that the individual members of a species can vary greatly in appearance and behavior. And he had realized that for natural selection to work, at least some of the differences among individuals that are being selected for or against must be passed to, and influence, the generations that follow. But what generates the differences in the first place?

Generally accepted mechanisms of the time, especially the idea of the inheritance of acquired characters that can be traced back to Aristotle and other ancient sources did not provide a mechanism by which this could happen. The mechanism that Darwin eventually proposed involved a kind of universal reproductive ability, which allowed all parts of the body to pass information about themselves to the reproductive cells and also to other parts of the body. In a book published nine years after the *Origin*, called *The Variation of Animals and Plants Under Domestication*, he proposed a mechanism for this kind of universal reproductive ability. He suggested that the information could be carried by microscopic mobile structures that he called gemmules. These tiny structures could migrate and bring information to new cells that were involved in the regeneration of body parts, or to the gonads during the formation of sperm and eggs.

Gemmules could change during an organism's lifetime, resulting in the inheritance of acquired characters. They could battle for the chance to be passed on to the organism's offspring. They could become inactive, preserving ancestral characters for generations, and then suddenly become active again. He proposed this last ability to explain how inherited traits could seemingly vanish and reappear in subsequent generations.

Pangenesis was a game try on Darwin's part. Alas, blood-transfusion (and presumably gemmule-transfusion) experiments between differently colored strains of rabbits, carried out by his cousin, the pioneering statistician Francis Galton, soon cast doubt on the idea [55].

Darwin was distressed and discouraged by Galton's results. But we can see now that his concept of inheritable particles of biological information did anticipate some of the recently discovered properties of the world of genetics to come. His gemmules were prescient glimpses of a world in which genes can sometimes move about on the chromosomes so that they are expressed differently in different parts of the body, or even jump between distantly related species and bridge billions of years of separate evolution.

Such jumps are rare. Most genetics is far more well-mannered. Mendel's paper detailing the laws of genetic inheritance in peas was published in 1866, but was only fully appreciated decades later. The Mendelian inheritance we are familiar with is far more orderly than the free-wheeling anarchy of gemmule fights or the acrobatics of jumping genes. And this ordered world is where I found myself as a student of genetics in the 1960s.

Watching Ecosystems Evolve

In 1963, when I was a beginning graduate student at Berkeley, I was given the chance to spend a summer getting hands-on experience with pools of genes. My task was to trap fruit flies for the pioneering evolutionary biologist Theodosius Dobzhansky (Figure 4.1).

Figure 4.1 Theodosius Dobzhansky (1900–75), one of the founders of the modern synthesis (photo by Don C. Young).

Dobzhansky, then based at Columbia University, loved to spend his summers working with his students in the national parks and forests of the American west. The trapping project formed part of one of his many long-running evolutionary projects.

Although I did not realize it at the time, this experience would shape my entire scientific career. Dobzhansky showed me how to follow evolutionary changes as they were happening in species living in the wild. I was to spend the rest of my life exploring the implications of his insights about how, and how quickly, the collections of genes carried by populations evolve. The insights of many scientists like Dobzhansky would lead to an understanding of how the world's evolutionary cauldrons boil and bubble.

That summer, Dobzhansky established himself in a spacious cabin at the Carnegie Institution's research station in the Sierra Nevada foothills north of Yosemite Park. Using the camp as a base, I traveled around the state, visiting remote areas where Dobzhansky's favorite fruit flies lived.

These flies were not the familiar *Drosophila melanogaster* that geneticists use in their laboratories. Dobzhansky jokingly referred to *D. melanogaster* as a "garbage species," because of its tendency to be attracted to garbage cans. The flies that he sent me out to trap, *Drosophila pseudoobscura*, were native to the western US, Canada, and Mexico, and he had been studying them for decades.

These flies were larger, darker, and, in Dobzhansky's view, altogether more impressive than the flies of that pathetic garbage species. Even before I arrived at the cabin, I had learned that these wild flies fully justified the sobriquet of "noble species" that Dobzhansky had bestowed on them. He had used their populations to get his early glimpse inside *gene pools*.

A gene pool is defined as the vast collection of genes possessed by all the members of an interbreeding population. The term can also be applied to the vaster pool of genes in an entire species. A gene pool is far larger than a *genome*, which is the set of genes carried by a single organism.

Dobzhansky, born in Ukraine in 1900 when it was part of the Russian Empire, had emigrated to the US as a young man. An accomplished linguist and a charismatic teacher, he attracted students from around the world. He was centrally involved in the establishment of thriving field and laboratory research programs in many parts of the world, including places as remote as the Brazilian rainforest.

In the 1930s, while a member of the famous "fly lab" that had been founded by Thomas Hunt Morgan at Columbia University and that later moved to CalTech, Dobzhansky teamed up with the cytogeneticist Calvin Bridges to probe the gene pool of his noble species.

Bridges had taken brilliant advantage of a developmental oddity found in the larvae of two-winged flies such as *Drosophila*. In the nuclei of the large cells of the salivary glands of their larvae, the thread-like gene-bearing chromosomes have the happy property of duplicating again and again during the cells' lifetimes, forming sausage-like giant chromosomes. A unique pattern of darkly stained bands along each of the chromosomes is clearly visible under an ordinary light microscope. The bands mark regions where different sequences of DNA and the proteins associated with them are located.

And, conveniently, the maternal and paternal copies of these giant chromosomes pair up along their lengths, so that any difference in the size or order of their bands is immediately visible.

The bands are not genes, but they can be used to locate the positions of genes. Bridges used the giant chromosome bands to figure out the locations of various large mutational changes. In one impressive example, he examined the chromosomal location of a mutation that changed the fly's eye from its normal oval into a bar shape. He could see that the mutation resulted from the duplication of a little sequence of bands on one of the chromosomes—a kind of genetic stutter that would have been quite invisible on ordinary, far tinier, chromosomes.

When Dobzhansky looked at the giant chromosomes in his noble species, he quickly found huge differences in the order of the bands along these chromosomes among different flies. Because the maternal and paternal chromosomes pair up, they can form complex systems of loops (Figure 4.2) [56].

These different patterns of bands along the chromosomes result from two mutations that each caused a break on a chromosome, flipping of the piece between the breaks, and rejoining of the broken ends. These events can produce *inversions*, in which the order of the bands (and therefore the order of the genes) in one region of the chromosome is reversed. The gene sequence ABCDEFGH might become AFEDCBGH. When the two chromosomes pair, this region forms a characteristic loop.

When Dobzhansky looked at the looped pairs of chromosomes for the first time, he realized that he was looking at inherited variation in a species' gene pool. He was looking not at the genes themselves, but at signposts on the paired chromosomes that there were in fact differences in the chromosomes' gene order.

When he and his students began to look at chromosomes from different areas where the flies lived, they found that flies in wet, high-altitude regions of the Pacific Northwest coast carried sets of chromosomes with one of two almost completely different collections of inversions from the rest of the species. In many of the matings between flies carrying these groups of inversions, the female offspring were fertile but the male offspring were sterile. Yet flies of the two groups looked identical.

Theoreticians had hypothesized that speciation must involve the separation of gene pools, so that eventually organisms belonging to one gene pool would no longer be able to exchange genes with those belonging to another.

Dobzhansky and the Harvard evolutionary biologist Ernst Mayr had predicted the existence of a variety of *isolating mechanisms* that could keep gene pools from exchanging genes. Perhaps, Dobzhansky thought, the different groups of chromosomal band arrangements that he and his many enthusiastic students were investigating could provide signposts to these isolating mechanisms.

Their chromosome surveys revealed something never seen before. Concealed in these populations of superficially identical flies, and invisible to the casual

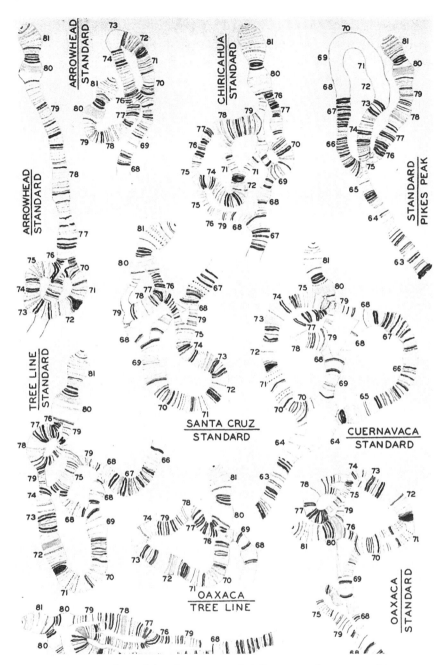

Figure 4.2 Loops formed by the pairing of different combinations of inversions carried by the third chromosome of *Drosophila pseudoobscura*. These configurations are easily seen in the giant salivary gland cells of its larvae.

observer, were two kinds of fly that were restricted in their ability to exchange genes freely with each other. Could they actually be different species, perhaps in an early stage of separation? And how would the discovery of such a seemingly highly similar pair of species impact the centuries-old definition of species as groups of animals or plants that are clearly different in appearance and behavior?

The differences in the gene arrangements of the two groups showed that there was little gene exchange between them. On the basis of this and other evidence, Dobzhansky promoted his second group of flies to the status of a separate "noble" species. His group had discovered a species that was in the process of being born—a species, as he memorably put it, *in statu nascendi*. He named this new *sibling* species *Drosophila persimilis*.

The two sibling species, *pseudoobscura* and *persimilis*, often live in the same forests. They had of course escaped notice earlier because they were almost indistinguishable in appearance and behavior. But, on careful examination, the males of the two species were found to have slightly different genitalia. They also vibrated their wings in slightly different ways, producing slightly different mating songs, and they preferred to pay court to females at different times of the day.

We now know that these species' ancestors began to diverge from each other about half a million years ago. We assume that they must now live in slightly different ecological niches, but we cannot be sure because the flies are shy. They spend much of their time living in secrecy somewhere in the forests, and only emerge when they are attracted by the alluring smell of overripe bananas.

Thus, most of the details of their mysterious real lives remain elusive. Some of their larvae have been found in oozing yeast-infected sap of oak trees. As Dobzhansky loved to say, "Ten percent of these noble flies live in oak tree slime fluxes. The remainder are produced by spontaneous generation!"

Many human genes are *polymorphic* (Greek for "many forms") for different *alleles* (variants of a gene), such as those that have given rise to our A, B, and O blood types. The fly populations that Dobzhansky was studying were *chromosomally polymorphic*.

Just as more than one blood type allele can be present simultaneously in a human gene pool, more than one different inversion can be present simultaneously in the gene pool of a *D. pseudoobscura* or *D. persimilis* population. Some of these inversions have been in one or the other of these gene pools for a long time, perhaps hundreds of thousands of years.

Dobzhansky and his students began to follow changes in the frequencies of these inversions, surveys that had extended over more than a decade when I began my California-wide sampling. Did the different chromosome inversions bind their genes into *supergenes*, able to perform slightly different tasks or to cooperate with each other in subtly different ways? Were these changes correlated with environmental factors, such as temperature and rainfall, or related to other factors in the flies' environment?

I quickly discovered that getting these data was much harder than it sounds. As Figure 4.2 shows, when a fly carries a pair of chromosomes that are separated by multiple overlapping inversions, the paired chromosomes often exhibit a nightmarish tangle of loops within loops. These nested loops can get squished into all kinds of rococo shapes when the chromosomes are prepared for viewing on a microscope slide.

I have vivid memories of peering through the cabin's microscope, using the light of a hissing hurricane lantern to illuminate and try to solve these three-dimensional puzzles. Often, and rather unnervingly, Dobzhansky would sit next to me, tapping his pencil and waiting for my verdict. Somehow, I managed to pass the tests. And, rather to my astonishment, I even got rather good at identifying these tiny glimpses into the flies' gene pools.

I was more than compensated for these periods of terror by wonderful candlelit dinners at which Dobzhansky would tell his student Wyatt Anderson and me vivid tales of the early days of genetics. We also talked at length about what these inverted pieces of chromosome might represent, but here we quickly found ourselves in deep water.

An inversion may happen to lock together sets of genes that confer an advantage to the flies. If the advantage of these locked genes is only expressed under some environmental conditions, and if under other conditions the genes are disadvantageous, the result can be a "balance" between conflicting selecting forces. Selective pressures that increase a chromosome's frequency might be balanced with those that decrease it. The point of equilibrium between these opposing forces might allow such a locked-together inversion to persist as a *balanced* polymorphism for many generations.

Dobzhansky, in the paper that we published on our summer's work [57], suggested that many of the changes in frequency we were seeing throughout California could be the result of environmental changes that were only starting to be understood, notably the effects of widespread pesticide use. But the shifts in frequencies that we were seeing were different in different parts of the state.

I now think that overall environmental changes in California could have played a role, but that it is also likely that many of the changes were happening because the flies were playing roles in different entangled banks in different parts of the state. Recent work on *pseudoobscura* inversions by Steve Schaefer's group at Pennsylvania State University has shown that different inversions carry different numbers of copies of olfactory sensor genes and detoxification genes [58]. These differences could have a wide range of possible causes—including increasing pesticide use.

Many things were clearly going on in these fly populations, but the details of what was happening were still mostly out of our reach. Even now, when we are able to apply our more recent ability to sequence large amounts of the DNA in gene pools, the stories that emerge can still be complicated and confusing. But some important things are now becoming clearer.

Features of gene pools that, back in the early 1960s, could only have been predicted as mathematical consequences of how genes ought to behave have now been shown to shape much of evolutionary change. New and unexpected discoveries about the amount of diversity that is concealed in gene pools, and about how the species carrying such diversity are interacting with each other, are coming thick and fast.

But before we explore this amazing new universe, let us turn to the new sequencing methods and the other molecular technologies themselves. Knowing a little bit about how these methods work will allow us to see more clearly how they are now giving investigators the power to probe the details of gene pools in ways that Dobzhansky, Wyatt, and I, working by lantern light in our log cabin, could not have imagined.

How to Sequence the DNA of an Entangled Bank

By the early 1950s, experiments with bacteria had already shown that DNA could act as if it were a gene. It could do so even after it had been painstakingly purified and freed of every trace of other types of molecule. Yet, its chemical structure was far simpler than that of proteins, which for decades had been assumed to be the genetic material. The DNA molecule was made up of large numbers of four smaller, negatively charged molecules or bases: A, T, G, and C.

Watson and Crick's discovery of DNA's amazing and elegant structure depended on beautiful pictures, taken by Rosalind Franklin [59], of the patterns that are produced when water-saturated molecules of DNA scatter beams of X-rays. These and other data, combined with leaps of intuition, led Watson and Crick to their famous model of a double helix with base A on one helix always pairing with base T on the other, and base G on one always pairing with base C on the other.

Watson and Crick saw that all that was needed to make a copy of DNA was to peel its double helix apart and synthesize new strands that are complementary to the old ones. The result would be two identical double helices, each carrying exactly the same information in the form of the sequences of bases along each of their helices. This process could repeat indefinitely. In their paper [60], they used one of the most breathtaking throwaway lines in all of the scientific literature: "It has not escaped our notice that the specific pairing we have postulated immediately suggests a possible copying mechanism for the genetic material."

With their masterful bit of understatement, and with ill-concealed glee, they signaled to their readers that they were indeed intelligent enough to realize that this postulated structure would solve the ancient riddle of the nature of genetic information and how it was inherited.

Five years after Watson and Crick's discovery, Marshall Nierenberg and his colleagues cracked the genetic code that DNA carries (and that the closely related single-stranded molecule RNA can also carry, though in slightly different form). In a series of papers they showed how the sequence of DNA's bases A, T, G, and C, when read in groups of three, determines the sequences of amino acids that make up the many proteins that are essential to the structure and survival of living organisms.

This was a pivotal moment. But it was only the beginning. It was as if we knew the letters of the alphabet, and knew that these letters were the key to the information in all the books of the world's great libraries, but found that most of the books were still locked up, like the precious illuminated manuscripts in a medieval monastery.

Learning how to read those books, and to understand even a tiny fraction of them, took another quarter of a century. The key figures were Fred Sanger at Cambridge University, who succeeded in accumulating two Nobel Prizes during his amazing career, and the duo of Allan Maxam and Walter Gilbert at Harvard. They developed slightly different methods for reading the base sequences of DNA directly. At first, the methods could only be used on short pieces of DNA that could be isolated and purified using the methods of the time, such as the tiny chromosomes of viruses and mitochondria.

Initially, the Maxam and Gilbert method was a little easier to use, but Sanger's ingenious method soon came to dominate the field of DNA sequencing. For years, scientists around the world, including those in my laboratory, passed strong electric currents through thin layers of a jelly-like material (a *gel*), to separate tiny fragments of radioactively labeled DNA according to their size. We carefully carried the glass plates that supported the gels into the darkroom. There, we positioned the plates on large sheets of film, allowing the DNA fragments to take pictures of themselves.

Sanger's sequencing method started with a short piece of single-stranded DNA and used the enzyme DNA polymerase to generate little complementary strands. Each such mixture was divided into four, and each of these was "poisoned" by adding small amounts of chemically modified A, T, G, or C bases that would halt the reaction when the polymerase enzyme added the poisoned base to the growing chain. The result was a collection of little fragments of DNA that were of different lengths, but that all ended with that "poison" base. Each of these collections of fragments was separated, according to fragment size, on its own lane of a gel, and the sequences of bands from each collection were compared with each other.

If a sequence of DNA were ATGGC, then there would be one small band in the poisoned A lane, a slightly larger one in the T lane, two successively larger ones in the G lane, and finally the largest in the C lane. The sequence of 100 or more consecutive bases in a DNA sequence could be "read" from a gel in each of these experiments.

The Maxam and Gilbert method was a little different, but it yielded exactly the same results if it was used on the same piece of DNA. In 1983, Martin Kreitman, then a student of Walter Gilbert and of the population geneticist Richard Lewontin, used the new Maxam and Gilbert method to probe part of a gene pool in detail [61].

Genomes are huge, so Marty was able to explore only the tiniest portion of one. He concentrated on reading the order of bases in a piece of DNA that was almost 2,700 bases long. Copies of this piece had been cloned and purified from each of eleven *Drosophila melanogaster* flies, a necessary step before sequencing. The stretch that Marty chose included the gene that coded for a well-studied enzyme, alcohol dehydrogenase, along with some non-coding regions.

Marty found that when he lined up these little DNA sequences, each comprising only 0.0015 percent of its fly's genome, there were base differences at forty-three points along their sequences. Some of the sequences carried one base or group of bases and others had a different base or bases. If such numerous differences were present in the rest of the flies' genomes, their collective gene pool must be filled with genetic variation.

Table 4.1 shows some of the most interesting variants that Marty found in his eleven sequences. For simplicity, the positions that had no variation among the eleven sequences are not shown.

Even with this limited amount of data, glimpses of the complexity of the *D. melanogaster* gene pool could be seen. The only one of the DNA differences that Marty found that actually had an effect on the amino acid sequence of the alcohol dehydrogenase protein itself was a genetic polymorphism that resulted from a single base difference (at position 1490 in Table 4.1). This tiny difference was enough to change one of the protein's 256 amino acids.

Half of the genes Marty sequenced carried an A, at this position, so that the code word of which it was a part specified the basic amino acid lysine. The other half of the sequences carried a C at the position, which changes the code

Table 4.1 This table shows a little portion of Marty Kreitman's DNA sequence data. The portion comprises a short, closely linked region around the one meaningful or non-synonymous change in the alcohol dehydrogenase gene at base position 1490. All the other differences in this particular segment are synonymous (changing the code but not the amino acid) or are in a region of DNA that is not translated into protein. Thus, these variants do not affect the amino acid sequence of the protein itself. One of the two possible sets of synonymous changes that might have been towed along by genetic hitchhiking to one of the bases at position 1490 is shaded darkly, and the set that might have been towed by the other base at 1490 is shaded more lightly (see text).

SEQ #	Position in the alcohol dehydrogenase sequence										
	1406	1426	1431	1443	1453	1490	1518	1528	1568	1597	1695
1	A	A	C	C	C	A lysine	C	T	A	G	C
2	A	A	C	C	C	A lys	C	T	A	G	C
3	T	C	T	C	C	A lys	C	T	A	A	C
4	T	C	T	C	C	A lys	C	T	A	A	C
5	T	C	T	C	C	A lys	C	T	A	G	A
6	T	C	T	C	T	A lys	T	T	C	A	A
7	T	C	T	G	T	C threonine	T	C	C	G	A
8	T	C	T	G	T	C thr	T	C	C	G	A
9	T	C	T	G	T	C thr	T	C	C	G	A
10	T	C	T	G	T	C thr	T	C	C	G	A
11	T	C	T	G	T	C thr	T	C	C	G	A

word to one for a very different amino acid, the water-loving amino acid threonine. Biochemical studies had already shown that both of these forms of the enzyme could function, but that each of the forms showed slight differences in its ability to carry out its task.

Marty was able to determine that all the other differences among his sequences had no measurable effect on the composition or function of the alcohol dehydrogenase protein. They were either in parts of the DNA sequence that were not located in the gene's code, or were changes that altered the code in a way that had no effect on the resulting protein. Nonetheless, these were genetic polymorphisms, adding to the flies' gene pool variation.

Such "silent" changes have been presumed to have little or no effect on the proteins, but many cases have since been found in which apparently "silent" changes do have an effect on the organism. They may, for example, alter how easily proteins can be translated or can fold into their active forms [62].

Some of these silent differences were puzzlingly arranged. The difference that Marty found at positions on the DNA sequence that were very near (*closely linked*) to the lysine-threonine coding difference seemed strangely to "echo" that meaningful coding difference. Almost always, at these nearby sites on the DNA sequence, one of the two silent bases was carried by the DNA sequences that also coded for the lysine amino acid, while the other silent base was carried by the sequences that coded for the threonine amino acid.

It was as if this little piece of chromosome was being inherited as a unit—even though the only one of these base polymorphisms that should have made any difference was the A-C polymorphism that codes for the lysine-threonine polymorphism.

Other polymorphic silent bases, which lie further away along the sequence from the lysine-threonine polymorphism, were also correlated with that polymorphism, but not as perfectly.

Even though Marty's data had included only a minuscule fraction of the flies' gene pool, they were already yielding an exceedingly detailed look at a piece of that gene pool's long history. My own much earlier counts of the numbers of Dobzhansky's inversions in different *D. pseudoobscura* populations, which I had painfully accumulated by peering through that lantern-lit microscope, were the equivalent of looking at a galactic nebula through Galileo's primitive telescope. The information revealed by Marty's sequences was like details of such a nebula seen through the James Webb telescope.

Marty had found forty-three genetic differences in his sample. More extensive data have been collected since, showing that if you compare any two sets of

D. melanogaster chromosomes point-by-point along their sequences of bases, about 1.5 percent of them are different. This adds up to a remarkable 2,700,000 tiny base differences between, for example, the chromosomes that a fly receives from its mother and those that it receives from its father.

It is likely that only a small fraction, probably less than 1 percent, of this variation actually makes a difference—positive or negative—in how the flies survive and reproduce. Still, there could be tens of thousands of genetic differences that really do make a difference.

What happens to all the differences in a gene pool, the meaningful ones and the trivial ones, as they are passed down through the generations? During each of these generations, these many thousands of differences get shuffled by *genetic recombination*, one of the most important of the processes that make evolution possible.

The enormous consequences of genetic recombination are often sadly neglected in discussions of evolution. In what follows, I will try to correct for this omission.

We must always remember that genetic recombination can only work its magic if there actually are genetic differences between the chromosomes that are being recombined. In Marty's fly population, there are clearly many such differences.

Recombination takes place during specialized cell divisions that lead to the formation of sperm and egg cells (*meiosis*). Each of these meiotic divisions exchanges pieces of each of the fly's pairs of maternal and paternal chromosomes, but each exchange involves different pieces, producing a different mix of genes in each of the sex cells that the fly makes during its lifetime.

In *D. melanogaster*, the males show few such breaks and rejoinings, while females can have a dozen or so. But the numbers are relatively small. In humans, the number of exchanges that take place in each of these cell divisions average about thirty. But in both species, the places on the chromosomes where the exchanges happen are different in each gamete-forming cell division. Thus, genetic recombinations between maternal and paternal chromosomes carrying many little differences ensure that all the gametes flies (or humans) produce are genetically different from each other.

Without sexual reproduction, and its attendant genetic recombination, our children would inherit a small number of huge blocks of genes, in the form of entire chromosomes, from their parents. And they would in turn pass on these blocks to their children. New variants of these blocks could only arise by occasional mutations, and the new mutations would stay only in the descendants of that block forever. Evolutionary change would slow to a crawl.

In contrast, genetic recombination can produce an unimaginable number of possible offspring, because it takes all the genetic differences between the maternal and paternal chromosome pairs that we each carry and shuffles them into astronomical numbers of different combinations.

In a gene pool as polymorphic as the *D. melanogaster* that Marty looked at, there is an explosion of genetic variability each generation. The genes themselves change seldom during this process, as a result of occasional mutations, during this process, but the potential number of ways in which the mutations that have accumulated can be recombined is immense.

Suppose that you, like one of Marty's fruit flies, carry an A on one of your maternal chromosomes and a C at the corresponding position on your equivalent paternal chromosome. Considering just this position, you could produce two types of gamete, one type that has a chromosome with an A at this position and one type that has a C. Nothing remarkable there.

However, elsewhere on the chromosome, or on a different chromosome, you might have a G on one chromosome and a T on the other. So, considering just the bases at these two positions, you could make four possible types of gametes: AC, AT, GC, and GT. Combinations of three different pairs of bases could yield eight possible gametes (two raised to the third power). Four could yield sixteen (two raised to the fourth power), and so on.

The maternal and paternal chromosome sets of the average human resemble each other more than those of the average fruit fly, differing in only about one base in every 1,000. But, because the human genome is much larger, a vast three billion bases, the chromosomes that we inherit from our mother and father differ at about three million bases on average. The number of possible gametes one of us could produce is therefore $2^{3,000,000}$, which, when converted to base 10 is about $10^{880,000}$, a 1 followed by almost a million zeros. Now, *that* is a large number! Our entire universe is a piker by comparison, containing a mere 2×10^{89} elementary particles!

Of course, even the most indefatigable of us produce an infinitesimal fraction of that number of gametes. And recall that the number of recombinational breaks and rejoinings during the production of each of our gametes is relatively low: about twenty-three in males and thirty-six in females. This means that we pass on to our offspring big chunks of our maternal and paternal chromosomes. These chunks carry large sets of genes that are still linked together.

Although an individual passes on a small set of such chunks into each of his or her gametes, all of the gametes carry a different set of chunks. Other members of the population are passing on different chunks of genes to each of their gametes. And the next generation scrambles everything again. These recombinations, even over only a few generations, quickly scramble a population's genes in an astronomical number of ways.

Recombination has immense effects, but note that genes near each other on a chromosome can "stick" to each other over many generations. It is as if the players in a poker game are eating candy. As the players shuffle the cards in the deck with their sticky hands, the cards that happen to be near each other will tend to stick to each other. This alters the odds of the game, because certain combinations of these sticky cards will tend to turn up more often than they should after each shuffle. And it is the "stickiness" of strong genetic linkage, persisting over thousands of generations, that may help to explain why Marty Kreitman's little clusters of alleles linked to that meaningful polymorphism have tended to stick together.

The brilliant work of Pardis Christine Sabeti [63] and of many others has shown that clusters of closely linked genes are preserved in a population because they happen to contain a genetic variant with a positive impact on their owners' fitnesses. As this useful variant increases in frequency in the population, unimportant or weakly selected genetic variants that happen to be closely linked to it can be dragged along with it. John Maynard Smith and John Haigh called this process "genetic hitch-hiking" [64].

The "silent" alleles that are closely linked to the polymorphism that has a substantial effect on the enzyme in Marty's sequences were probably carried along by hitchhiking to one of those code-changing alleles as that allele swept to high frequency in a fly population in the past.

Why does each of the protein-changing alleles in Marty's data set have different sets of neutral-appearing polymorphic alleles closely linked to it? Perhaps one of these protein-changing alleles was introduced into the gene pool of these flies from a different population of *D. melanogaster*—one that has not yet been discovered or that may no longer exist. If that second allele happened also to carry a set of closely linked silent alleles that were different from those linked to the first allele, all those differences could come along for the ride as this introduced selected allele spread through the new population.

All this, of course, is to some degree guesswork. But it is guesswork based on real, incredibly detailed data. During the third of a century since Marty's pioneering work, almost ten trillion bases of DNA, from half a million species, have been sequenced. We can ask—and answer—questions that could not have been imagined by previous generations of scientists.

And gaining this data has advanced far beyond the "dripping gel" period that we all suffered through during the early stages of sequencing. As of the time of writing, a 1,000-dollar investment and a few hours of time on a sequencing machine can yield a gigantic bolus of genetic information that is thirty-five million times as large as Marty Kreitman's painfully acquired glimpse of one small drop in a gene pool.

Now, if you use an advanced sequencing method that feeds single DNA molecules through tiny pores in a membrane and reads the sequence of bases as they pass through, your mass of data can include some stretches of DNA molecules that are as much as four million bases long.

We can now fish out, even from DNA in the air or water, information about the gene pools of healthy, endangered, and even previously unknown species, ranging all the way from the largest to the smallest inhabitants of ecosystems. Recently, fragmentary but still usable information has been obtained from two-million-year-old estuarine mud taken from beneath a currently glacier-covered region in northern Greenland. The preserved DNA provides a picture of a thriving open forest ecosystem that teemed with mammoths and other large mammals [65].

We are learning what kinds of genetic variation these gene pools contain. We are also becoming able to see the patterns that result when living cells modify their DNA and RNA bases in ways that change how their genes function (though, because these changes are temporary, they do not affect the genetic sequences themselves). And we are beginning to glimpse consistent patterns that serve as warning signals about species extinction and ecosystem damage.

The new technologies that we have explored here, and others that we will encounter later, can reveal not only the evolutionary cauldrons' contents but also their enormous potential and their ability to enhance the process of evolution itself.

5

How Entangled Is an Entangled Bank?

So the vast number of species that exist today, which we still have not completely catalogued, has come into existence. Its complexity, its vast web of interrelationships and dependencies, is almost beyond comprehension. So complex is it, that we cannot predict with any certainty or accuracy the effects of damage to any one part of it. But that complexity is something we must do our utmost to protect, for it is that which enables it to absorb the worst effects of damage and to heal itself.

David Attenborough, *Life on Earth* (1979)

The new universe of information that has been revealed by DNA sequencing of the gene pool of a single species such as *Drosophila melanogaster* or *Homo sapiens* is dwarfed by the multiverses of information that are contained in even the simplest ecosystems. In this chapter, we will explore some of these multiverses and the challenges that they present to the intrepid molecular ecologists who are beginning to probe them. We will discover that Darwin's entangled banks are far more entangled, and are even more central to the process of evolution, than he could have imagined.

Ecologists have long assumed that the ecosystems of the open ocean, unlike those of coral reefs, must be extremely simple. These enormous, continually mixing bodies of seawater are illuminated near their surfaces by sunlight that diminishes rapidly and falls to insignificant levels below about 200 m. Below the limit of light penetration lies Stygian blackness and near-frigid temperatures.

Whatever life these waters can support is often limited by shortages of organic molecules, phosphates, and iron. The ecosystems of tropical and subtropical oceans, which are only rarely influenced by the upwellings of colder mineral-rich water that occur nearer the coasts, seem to have so little life that they have been called oceanic deserts.

I admit to having little direct experience with such alien ecosystems. But in 2013, I was on an oceanographic trip to the tiny unpopulated coral atoll of Clipperton, named after an eighteenth-century pirate. This isolated speck of land lies 1,500 km southwest of the southern tip of Mexico's Baja California peninsula.

We were about a day's sail from the island when the captain suddenly announced that the ocean beneath our keel was three kilometers deep. Why not, he suggested, go for a swim?

The waters were calm, and with the engines turned off the boat soon lost headway. How often does one get a chance to swim in a gigantic swimming pool that is three kilometers deep? All of us eagerly lined up to get wet. I jumped into the clear, warmish water without a wetsuit, and peered through my facemask into the translucent green world beneath the ship. Nothing but green depths greeted me—no fish or other small organisms were visible.

As I was swimming about, looking for creatures to photograph, I suddenly felt a searing pain along my left upper arm. Whatever had caused it was not apparent. I called to everyone to alert them, and swam hastily toward the ship's ladder.

It was likely that a small jellyfish had stung me, one that was so tiny and transparent that I hadn't been able to see it. Jellyfish, and other oceanic drifters such as the Portuguese man-of-war (a colonial coelenterate that is a remote relative of the corals), are becoming more common as the oceans warm. Luckily, there was a dermatologist on board our ship, and he treated me with a strong steroid ointment that worked wonders.

Even this remote, apparently simple ecosystem, unable to support much visible life, was still capable of harboring some painful surprises. But even as I was splashing about in that seemingly featureless oceanic world, evidence was rapidly accumulating that such ecosystems are anything but simple.

You will recall the classic experiments of G.F. Gause and A.C. Crombie, who used *Paramecium* and flour beetles to investigate competition between species (Chapter 1). Their experiments strongly supported one of the earliest laws of ecology, the *principle of competitive exclusion*. The principle states that in a simple ecosystem with a single food source and no environmental complexity, a more successful competitor will drive out a species that is less successful.

Surely, such ecological simplicity should apply to ecosystems like those of the deep ocean. These ecosystems are dominated by one-celled photosynthetic phytoplankton that spend most of their time drifting or swimming slowly in the sunlit upper waters.

The entire planet depends on these plankton. These little single-celled plants produce more than 50 percent of the world's free oxygen, even though they make up only about 1 percent of the total biomass of photosynthetic organisms. These cells are so productive because they are all business.

Larger plants, such as the trees that dominate many land ecosystems, are mostly made up of supporting tissues that are needed in order to hold their leaves and fronds up to the light. Such plants must also grow immense root systems, which anchor them and allow them to absorb water and minerals from the soil.

The tiny phytoplankton cells, as they float and swim freely, need only follow the light and photosynthesize. They are surrounded by all the water they might require. Seawater also provides them with any needed minerals. Even though some of these tiny creatures can have animal-like characteristics, such as tiny flagella that enable them to swim about, they do not obviously fight with each other or stake out territories. They simply use sunlight for the energy that they need in order to build organic compounds and to pull essential minerals out of the surrounding water.

Surely, therefore, only the best photosynthesizers and mineral-pullers, those that are most able to use these simple resources to out-reproduce their competitors, should prevail in such systems. But samples of plankton taken by early oceanographic expeditions showed that many clearly different phytoplankton species, drawn from many branches of microscopic life, live in even a small sample of midocean seawater.

These oceanic ecosystems are also able to support many different species of photosynthetic and non-photosynthetic bacteria. Some of these bacteria just need sunlight, carbon dioxide, water, and a few simple elements and compounds to build everything that they need. Others require a supply of organic molecules from the environment in order to survive.

Why, in what is surely a simple ecosystem, is there so much diversity? Ecologist G. Evelyn Hutchinson confronted this puzzle in a 1962 paper, "The Paradox of the Plankton" [66]. He looked at the distribution of abundances of phytoplankton species that were known at the time and concluded that this distribution could not be explained as the result of direct competition among species. Common species were too common, and rare species too numerous, to be accounted for by simple ecological models.

Since that time, new revelations about the marine world's biological complexity continue to arrive at a brisk and sometimes overwhelming rate. Additional levels of complexity in the oceans' physical environment also continue to be revealed, but at a much slower pace. Taken together, these observations are consistent with Darwin's view that between-species interactions, not direct interactions with the physical environment, are normally the primary drivers of ecosystem evolution.

A Cornucopia of Diversity

An unforgettable image, bequeathed to us by the ancient Greeks, is that of the cornucopia, which appears in various legends in the shape of a horn broken off from the head of a ram or possibly the head of a god. This mythic cornucopia pours forth an endless stream of plenty.

Molecular biologists and physical chemists are now able to use nature's richness and their own enormous ingenuity to tap into a molecular cornucopia of information about the gene pools of individual species and of entire ecosystems. What follows is a brief summary of these riches, and of how we are beginning to glimpse their full extent.

All living cells, as well as the viruses that parasitize them, carry two great libraries of molecular information about themselves. One library is made up of chemically stable double-stranded DNA. The other is made up of its close relative, the more chemically fragile single-stranded RNA. Both types of molecule are able to carry genetic information down through the generations, either directly as the coded information itself or in the form of strands with base sequences that carry the information's complement and that can subsequently be used by replication enzymes as a template to recreate the coding molecule itself.

The enzymes that copy this information-transfer system can sometimes make errors that result in mutational changes in the coded information. Chemicals, ultraviolet light, and high-energy radiation can also produce mutational changes.

Without such mutations, evolution would come to a halt. But if these changes happen at too high a rate, the gene pool of a population might fill up with harmful mutations. Surviving members of the population would be at an ever-growing disadvantage. We have evolved DNA repair and other mechanisms that blunt the onslaught of harmful changes.

Populations with no ability to shuffle or exchange genes would be stuck, as every member of these populations would accumulate a growing number of harmful mutations—an inexorable downward trend that geneticist Hermann Muller called a "ratchet." Each harmful mutation is a click of the ratchet, causing the non-recombining population to go down, but not back up.

Recombination affords an escape from the ratchet. Some members of a recombining population are sure to have fewer than average new harmful mutations and higher fitness.

The organism's complexity and lifespan also play a role. Mutations are likely to have a more negative impact on complicated, long-lived, and slowly reproducing organisms than on the far simpler and far more fecund bacteria. It is therefore no surprise that our own species carries an especially expert set of DNA repair enzymes.

Because of such repairs, and because of chance events that often lead to the loss of mutations while they are still rare, most mutations, even beneficial ones, are lost before they can spread in gene pools. Luckily for the evolution of life on our planet, some beneficial mutations occasionally survive and spread.

Interestingly, there is a subtle evolutionary benefit to being a bacterium and having a much less effective set of repair enzymes. Most bacteria are haploid—that is, they have only one set of chromosomes (unlike our diploid selves with one set from each parent). In diploids, mutations are often recessive, masked by the functional allele from the other parent. But the advantage of a beneficial mutation is likely to be fully expressed in a bacterial cell as soon as the mutation happens. This immediate effect enormously increases the likelihood that a beneficial mutation will spread in a haploid bacterial population compared to a population of big, slowly reproducing diploid organisms such as ourselves. Similarly, bacteria will quickly lose harmful mutations because their harm is not masked by a functional copy of the gene.

High mutation rates allow bacterial populations to adapt quickly to new environmental conditions. The cost in cell death may be high, but bacteria usually reproduce so quickly that their populations can easily survive even a steep cost. We will shortly see a vivid example of such bacterial resilience.

Probing Microbiomes

We now have many tools that allow us to probe not just gene pools, but entire sets of gene pools. In addition to a growing armamentarium of tools for replicating DNA and RNA, molecular biologists have discovered a huge variety of editing enzymes that can cut DNA or RNA into fragments and join them together in new patterns. These enzymes have been isolated from bacteria, from plants, from mammals like us, and from many other organisms.

Other enzymes are available to carry out pretty much every manipulation of DNA or RNA that you can think of—and more besides. DNA-manipulating enzymes can be harnessed to perform a polymerase "chain reaction," which can make enormous numbers of copies of any pieces of DNA or RNA. These multiplied copies can be sequenced using yet other kinds of enzyme.

These and many other molecular tools can probe both our own genomes and the genetic and ecological cornucopias that surround us. Employing these techniques, in 1998 Jo Handelsman and her colleagues invented and named the first—and still enormously effective—approach to tackling the nature of genetic diversity in an entire ecosystem. They called their method *metagenomic analysis* [67].

Even a tiny sample of sediment, dirt, or dust from the environment can contain a huge number of DNA molecules. Avogadro's number tells us why.

This number is defined as the number of individual molecules of an element or compound that are present in an amount of that element or compound that is equal to its molecular weight in grams. For example, the hydrogen gas molecule is made up of two atoms of hydrogen, the lightest element. The molecular weight of each atom of hydrogen is defined as one, so the molecular weight of a molecule of hydrogen gas is two. But the number of such molecules in two *grams* of hydrogen, a mere whiff of gas, is immense, totaling 6.23×10^{23}. This is Avogadro's number.

This number remains constant regardless of the size of the molecule. Of course, the amount of a molecule needed to equal Avogadro's number will be different for different molecules. You would need a kilogram of a compound with a molecular weight of 1,000 in order for it to contain Avogadro's number of molecules of that compound.

DNA molecules are even more massive. But, because Avogadro's number is so huge, even a tiny amount of DNA can contain a lot of molecules. One millionth of a gram of DNA molecules, each with a molecular weight of one million, still contains 600 billion molecules.

As a chemist friend once said to me, "Avogadro's number is a very big number."

Metagenomic analyses starts with DNA purified from a sample of water, soil, or even dust that floats in the air. The sample may contain DNA from many thousand species. Avogadro's number means that when this mix of molecules is replicated by the polymerase chain reaction to make more copies, and the resulting DNA fragments are sequenced to produce an immense jigsaw puzzle, the puzzle is sure to contain pieces of the genomes of many of these species.

Computer programs can search and sort the sequenced pieces of DNA. Such searches, which were initially used to speed up the Human Genome Project, begin by pairing up the fragments that share part of their sequence. The program then looks for other pieces that share part of this new longer sequence. It adds them to the sequence, building longer and longer sequences.

The entire genomes of bacteria and viruses, which have small chromosomes, can often be reconstructed from amplified pools of DNA fragments taken from an infected individual, from the individual's environment—or even, as we will see, from a single living cell or virus. For larger organisms, matches to genes that are already in the database are usually enough to identify which fish, or insect, or worm the fragment likely belongs to—or at least what its close relatives are. And sometimes the pool of bits of sequence is large enough, and the bits are long enough, to reconstruct whole chromosomes that belong to such a large organism and even its entire genome.

The number of organisms that can be detected in just one sample of sea-water, soil, or even dust from the air usually dwarfs manyfold the number of species that Darwin was able to study during his voyage.

But there are three problems. First, metagenome analysis can yield the genomic sequences of many of the organisms that are present in a sample of soil or ocean water, or the teeming creatures that live in our digestive tract. Unfortunately, until recently only a few percent of these organisms could be cultured in the laboratory. The entire genomic sequence of an organism might be known from metagenomics, but if the organism itself is unculturable, many of its properties remain unknown.

Recently, this unculturable fraction of the microorganisms living in our own bodies has been reduced from more than 90 percent to about 60 percent through the use of clever culture methods [68]. But such methods have not yet been adapted for use in samples from the soil or from the marine world.

Most of the genome sequences that have been pieced together from metagenome analyses of midocean environments belong to organisms that are readily able to survive and play a role in their seemingly boring and featureless worlds, but that spurn the blandishments offered by cozy Petri dishes and delicious laboratory media. The things that these unculturable creatures do in their ecosystem are largely unknown to us, and yet they manage to thrive there.

The second problem is that there are often differences in DNA sequence between otherwise very similar sequences that are found in pools of metagenome fragments. Are these differences simply sequencing errors? Are they small and inconsequential differences between individual genomes, like most of the genetic variation that Marty Kreitman uncovered? Or do they show that the sequences actually come from slightly different species that, like *D. pseudoobscura* and *D. persimilis*, should not be lumped together as if they were a single species? And if some of these different sequences really are from different species, how many and what kind of genetic differences between them might we need to find in order for us to declare that their sources really are different?

Indeed, what exactly is a species of bacterium or virus anyway? These organisms mostly reproduce by making exact copies of themselves, and only occasionally exchange genes through various kinds of genetic recombination and gene exchange. Perhaps we should think of them as collections of clones rather than species.

Or perhaps not. Bacteria and their viruses have evolved an extensive array of crude and clumsy approximations to sexual reproduction, through various sex-like interactions and the effects of a wide variety of gene transfer agents. Such recombination events happen irregularly, but they can sometimes even involve the transfer of genetic information from one major branch of the tree of life to another.

In the microbial world, more commonly than in our own much more circumscribed reproductive world, between-species transfers provide a way of spreading the effects of advantageous genes among many different competing genomes. This could not happen if the genes of a bacterial species were locked into that particular asexually reproducing lineage and were thus all forced to share that lineage's fate.

Our growing understanding of the prevalence and mechanisms of these cross-species transfers, which geneticists lump together under the term *horizontal gene transfers*, gives us the opportunity to test the evolutionary processes that are at work in the entangled ecosystems that have formed a central theme of this book. Unfortunately, most of the experiments on such mechanisms so far have been carried out on pure lines of bacteria and on particular agents of pseudo-sexual reproduction, all grown under uniform laboratory conditions.

Yet, as we have seen again and again, in real entangled ecosystems there are no pure lines or constant growth conditions and there is a zoo of potential gene transfer agents. If we are to understand the true extent of the evolutionary changes that can take place in entangled banks, we will have to recreate an approximation of natural conditions in the laboratory.

Every bacterial gene pool in real entangled ecosystems is a mixture of different genotypes. These bacteria are all playing complex roles in ecosystems as they interact with the many other bacterial species, with the clouds of viruses that drift among them and those that lurk in their cells, and with the more complicated animal and plant hosts that they infect or interact with mutualistically.

The roles that gene transfer systems play in evolution are just beginning to be explored. I am going to take a leap of faith and predict that such studies will tell us much about the world of entangled ecosystems and about the evolutionary opportunities that these entanglements are continually providing.

The sheer number of differences between microorganisms is obviously important, but so are the kinds of differences. If most of these differences are untranslated or synonymous DNA changes, like most of the polymorphisms that Marty Kreitman had found in his *Drosophila* DNA sample, we might be hesitant to promote each of these slightly different groups of bacteria and viruses to their very own eminences as separate species. But if many of the differences have effects on the proteins that they code for, this might make an argument for bestowing the title of species on them more convincing.

Finally, who lives where? Viruses spend part of their life cycles inside the cells of their hosts, often co-opting their hosts' metabolisms as they replicate. A virulent viral infection causes the host cells to break open, releasing clouds of virus along with the hosts' own inner store of proteins and nucleic acids. But sometimes the viruses may persist inside their hosts for generations.

Which species of host does a particular virus inhabit? How many types of virus can only live in a single species of host, and how many are versatile enough to live in more than one host species? How do the viruses interact with their hosts, in addition to infesting and sometimes killing them? Crude metagenomic analysis, which pools all the genetic information of an ecosystem together into one single giant *gemisch*, is silent on all of these questions.

Now, new techniques of single-cell genomics are being used to find some answers to all three of these questions. They can pinpoint especially numerous or unusually distributed species that will repay further study in the laboratory. They can yield data on the differences between microbial gene pools, and on whether and how often they exchange genetic information. And they can peer into single cells to see how they interact with their parasites and possible symbionts.

Single-cell genomic sequencing of individual cells or viruses from a sample of seawater begins with some fancy hydraulics. A stream of liquid from a tiny sample of the seawater is broken up into tiny droplets, some of which contain a single virus or bacterium. UV light and chemicals are used to make these cells or viruses fluoresce. The droplets that contain fluorescing cells are then directed by electrostatic forces to tiny chambers in a "laboratory on a chip."

Within each of these chambers, proteins and other molecules of the cell or virus are stripped away, leaving only DNA or RNA. Copies of pieces of this DNA or RNA are then generated by repeated rounds of enzyme-driven replication. The resulting collection of replicated fragments from each droplet is sequenced to yield the droplet's own tiny molecular jigsaw puzzle.

The pieces of this puzzle are drawn not from an entire ecosystem, as they are in the analysis of a whole metagenome, but rather from that single bacterium or virus taken from the ecosystem. When computers fit the pieces of the puzzle together, the result does not yield, as metagenome analysis does, a glimpse of the genomes of an entire collection of organisms. Instead, it yields the complete sequence of the genome of the single cell or virus that ended up in the tiny chamber. And the sequencing might yield the genome of a host cell along with those of the viruses that it contains.

The techniques of single-cell and single-virus genomics were first used in studies of tumor biology [69]. They have permitted genetic changes in tumors to be studied at the level of their individual cells. Evolutionary ecologists have been quick to apply these advances to single cells isolated from entire ecosystems [70]. Early results, taken from seemingly featureless marine ecosystems similar to the one that I swam through in the middle of the Pacific, are now becoming available. They open up astonishing details that had previously been beyond our reach. But first, suitable samples had to be obtained from far-flung ecosystems.

The Voyages of the *Tara*

In 2009, the French ocean-going sailing schooner *Tara* set out from its home port of Lorient on the southern coast of Brittany. It was about to begin a three-year voyage of discovery funded by the redoubtable designer and philanthropist Agnès Troublé, the founder of the French fashion house *agnès b*. By the time the schooner returned in 2012, it had traversed 125,000 km in the Atlantic,

Pacific, Antarctic, and Indian Oceans. During the voyage, its scientists and crew collected 35,000 beautifully preserved and carefully curated seawater samples from shallow and midocean depths at multiple sites. Subsequent *Tara* expeditions have covered the Mediterranean basin and the rich Coral Triangle of Southeast Asia.

The *Tara* is tiny. At a length of 36 m, it is just a few meters longer than Darwin's *Beagle*. It was originally named the *Seamaster*, and already had a long history as a research vessel. Its first owner, the New Zealand yachtsman Peter Blake, had used it to make surveys of Antarctica and the Amazon basin. The scientists on board had searched for early signs of global warming and environmental degradation.

Then, in the vast mouth of the Amazon in 2001, tragedy struck. The boat was boarded by pirates, who shot Blake and killed him.

Today, the rechristened *Tara* is continuing to explore the oceans. Its new life as part of the *Tara* Oceans project is far from a vanity project.

Agnès Troublé also comes from a seafaring family, one that has taken part in America's Cup races. She has fully supported the use of her ship as an oceanographic vessel, and has been involved in some of the many educational projects in schools around the world that introduce students to the *Tara*'s travels. She has been able to encourage government policy initiatives that have been sparked by the vessel's findings.

Chris Bowler of the *École Normale Supérieure* in Paris was one of the scientists on board during *Tara*'s first oceanographic voyage. He told me recently of his many adventures, including brief encounters with pirates in the Philippines and the Mozambique Channel.

By mutual agreement, the crew of *Tara* had been unarmed since the Amazon boarding. They were also alert to suspicious activity. As soon as they spotted anything unusual in their vicinity, they obeyed the dictum that discretion is the better part of valor, stopped their sampling activities, and sailed away.

Although *Tara* is tiny compared to most modern research vessels, it is far more efficient at its task. It is able to catch a favorable wind to save on fuel, and can operate with one-fiftieth of the daily budget of a large ship. The ship's crew and its scientist passengers work as a team, and the *Tara*'s small draft allows them to reach regions that are off-limits to bigger ships.

Even large, stable oceanographic vessels can be tossed around by huge waves that are often encountered far from land. Because of the *Tara*'s small size, the weather posed an even bigger challenge. But the scientists and crew were still able to centrifuge samples carefully and group them into size categories that ranged from the tiniest viruses up to zooplankton that were as much as 2 mm long. The larger organisms were documented in millions of photographs yielding detailed three-dimensional images. All of the organisms were preserved under conditions ensuring that their DNA and RNA were undamaged.

Scientists from around the world have been given free access to the samples. The result has been a blizzard of papers in top scientific journals that reveal the planktonic world in unprecedented detail.

Virus DNA was found in each of the size fractions in the samples, demonstrating that viruses are indeed everywhere. Perhaps the biggest surprise to emerge from these first studies was how many different kinds of virus there are in plankton communities.

Most of these oceanic viruses are *bacteriophages* or bacteria-eaters, while some inhabit the cells of organisms larger than bacteria. Metagenomic analysis from the first expedition's data revealed forty million distinctly different sequences among the genomes of these DNA viruses.

When the oceanic bacteriophages attack their host bacteria, they use their hosts' cellular machinery to make more bacteriophages. These offspring break open the host's cells and float away. The bacteriophage populations are so dense and voracious that each day they destroy a third of the bacteria living in these communities. This immense mortality is the microbial equivalent of the toll that was exacted on the human population of Europe from 1347 to 1351 by the plague bacilli of the Black Death. But the bacterial slaughter happens daily! Luckily, since bacteria are so prolific, they can replace those that have been lost within hours.

The Black Death had evolutionary consequences. It brought about significant changes in the gene pools of European populations. For example, more than 200 immune-related genes changed in frequency, moving in the same direction, in both British and Danish populations during the 1347–51 plague outbreak. The shifts could be seen in DNA samples from graveyards of the time [71]. It seems likely that evolutionary changes must be taking place even more rapidly in bacterial populations that are subject to daily Black-Death-like pandemics, especially when species in these ecosystems can transfer genes to other species.

Lysis of bacterium cells by bacteriophages makes the contents of their destroyed hosts available to feed other species of bacteria, along with many other types of plankton that depend on the released goodies. This process of transfer of energy from one group of microorganisms to another has been named the *viral shunt* [72].

This shunt, only recently discovered, turns out to play a central role in the recycling of nutrients in these watery environments. Without the shunt, huge numbers of dead photobacteria would slowly sink below the reach of light, taking their precious nutrients with them to the bottom of the ocean.

Many oceanic bacteriophages are *temperate* rather than lytic. They lurk in their hosts' cells until, perhaps after many host generations, environmental triggers set them to multiplying and killing their hosts. Sometimes they *integrate* their chromosome into the host's chromosome. There they can lurk permanently, like a time bomb, or even entirely lose their ability to free themselves and eat their host. Such permanent integrations provide sources of additional genes to the bacteria.

Viral shunts, and the many other interactions that help to maintain the function of an ecosystem, are not the result of natural selection acting on the entire ecosystem as a unit. Instead, they are the result of a potpourri of distinct selective forces acting on the members of different species. Even though the balance of these forces is continually shifting, the shunt has emerged and been maintained.

Such natural selection acts on individuals, changing the gene pools of their species. It selects for those individuals best able to seize the opportunities that their environment provides. And those opportunities may—or may not— select for species that perform a particular function within an ecosystem especially well.

Suppose the opportunities presented to individuals of a species consist of the chance to parasitize another species that is important to the functioning of an ecosystem. Then, no matter how central to that functioning the other species may be, this destructive opportunity is sure to be seized. We will shortly see a vivid example of such seemingly counterintuitive selective opportunities in a simple laboratory ecosystem.

Complex ecosystems are protected against such damage by their very complexity. For example, if many bacterial species are present in an ecosystem, many ecological niches become available for different and perhaps new types of viruses that can prey on them. Bacteria that evolve resistance to these viruses will gain a short-term advantage, but new virus strains soon evolve that can overcome these resistances. These arms races are never-ending and can take different paths in different species of bacteria and virus. The complexity of such an ecosystem insulates it against the emergence of potentially disastrous imbalances that could jeopardize the existence of centrally important species.

Simple as many marine and freshwater ecosystems might at first glance appear, they really are diverse. They provide numerous opportunities for inter-species interactions that can drive the ecosystem toward even greater diversity. Complex Darwinian entangled banks can readily evolve even in such seemingly featureless places. And extensive diversity is exactly what the *Tara* surveys revealed.

A Glimpse of a Marine Entangled Bank

Using *Tara* samples from the western Mediterranean, researchers followed up on an intriguing discovery about the commonest type of bacteria, and the bac-teriophages that prey on them, in these deep-water ecosystems [73].

The commonest marine bacteria throughout all of the Earth's oceans are tiny little curved rods that are grouped together under the tentative name of *Pelagibacter*, meaning "bacterium floating in the open sea." These bacteria are not photosynthetic, but they can still synthesize most of the compounds that they need from simple molecules. Their evolutionary branch has been traced back to a common origin with related ancient bacteria that invaded the cells of our own distant ancestors and subsequently evolved into our own cellular powerhouses, the mitochondria.

Pelagibacter are difficult to culture in the laboratory, which means that they fooled investigators into thinking that they are rare—that is, until their DNA began to turn up so often in metagenomic analyses, that it belonged to one in three of the bacteria in plankton samples. Because they can easily switch between high- and low-oxygen worlds, they are present in abundance even at depths where there is little free oxygen.

Therefore, *Pelagibacter* surely play a huge role in the ocean's chemistry. The role its ancestors played might have been even more important during past times, such as the period around the great Permo-Triassic extinction a quarter of a billion years ago. During this time of upheaval, triggered by vast volcanic eruptions, the oceans became largely anoxic.

Pelagibacter is full of bacteriophages. One of them—unsurprisingly, in view of the abundance of its hosts—turns out to be the commonest virus in the oceans. This bacteriophage rejoices in the name of vSAG-37-F6. Like its host bacteria, it was for a long time largely overlooked in oceanic surveys even though its commonest protein is easily detected in ocean water samples.

And, until the advent of single-cell genomics, there were no clues about which bacteria these viruses preyed on or how many of them there really were.

These viruses and their hosts are so common that their interaction plays a central part in the viral shunt. As the vSAG-37-F6 virus grows and multiplies, it kills enormous numbers of its host *Pelagibacter*. Its activities fill the water with bacterial remains, along with a huge floating population of its own progeny.

This slaughter ensures that *Pelagibacter* does not take over the oceans. Instead, a variety of other bacterial species can use these floating building blocks to multiply. These bacteria, in turn, provide a smörgåsbord of food sources for a variety of other organisms farther up the food chain—and for other bacteriophages.

Single-cell genomics allows investigators to discover just how many distinct genetic variants are concealed in these bacterial and virus populations. The largest such survey to date, carried out on several deep- and shallow-water samples taken from western Mediterranean sites during the *Tara* voyage and other voyages, showed a high diversity of strains in both the bacteria and the vSAG-37-F6 viruses that prey on them.

In making sense of these observations, the investigators had to wrestle with the species problem that confronts all workers in the microbial world. There are indeed many genetic variants within populations of bacteria and viruses. Are they simply variants that constitute mere details of a single giant gene pool? Or are they signals that the bacterial and virus populations are divided up into different gene pools that exchange few or no genes?

Theodosius Dobzhansky was confronted with the same question when he made the momentous decision to divide his *D. pseudoobscura* populations into two separate species. He concluded that, judging from the fact that few of the inversions in the flies' giant chromosomes were shared between the two groups of flies, these two groups really do exchange very few genes at the present time. But his decision was largely a leap of faith, because he had only a tiny bit of information about the flies' gene pools.

Bacteria and viruses, in contrast to fruit flies, primarily reproduce asexually. They do occasionally exchange genes by mechanisms that lead to genetic recombination, but these exchanges tend to be ad hoc. They are not a basic part of their life cycles, like the recombinations between our maternal and paternal chromosomes that our gamete-producing cells perform each generation.

In the *Tara* samples many of the strains, both of bacteria and viruses, were so genetically different from each other that their DNA sequences shared only 95 percent or fewer of their bases. How does this compare with differences between known pairs of species?

When the DNA sequences of humans and chimpanzees are lined up, they share 96 percent of their bases. Thus, by this measure, the most similar of these bacterial and virus strains are as similar as humans and chimpanzees. Further, just as with humans and chimpanzees, there are usually no signs that these similar bacterial or virus strains have recently exchanged parts of their genomes.

So, can we call these bacterial and virus strains different species? The *Tara* scientists adopted a kind of extension of the *biological species concept* that Dobzhansky, Ernst Mayr, and their collaborators had introduced back in the mid-twentieth century.

If two groups of these microorganisms differ by 5 percent or more of their bases, and if they show little or no sign of recent genetic recombination between them, then we can take a deep breath, cross our fingers, and consider them to be different biological species just like humans and chimpanzees. This definition can be applied regardless of how the organisms reproduce, or whether they even show recognizable mating behaviors.

Classically trained biologists may (and do) gnash their teeth over such a permissive definition of species, but regardless of its overall merits it has the advantage that it can be applied consistently in many situations, even in planktonic communities. When the study's authors did this, they found that worldwide, the *Pelagibacter* host bacteria could be grouped into an astonishing 495 species. Of these, fourteen were found in the samples from a small part of the Mediterranean that they were examining in their single-cell genomics study.

At least some of the cells from five of these bacterial species were infected by viruses. When the authors used the inclusive species definition on those viruses, they could detect 123 virus "species." Remember that this remarkable virus diversity was discovered in that handful of samples of water from a small region of the Mediterranean. And note that all the other bacterial and virus species present were ignored in the study.

The authors were not as certain of the virus species as they were of the bacterial ones. Virus genomes are much smaller than bacterial ones, so that the genetic differences among the virus strains are far fewer. But the researchers strongly suspected that their classification was indeed reflective of real underlying genetic differences that separated those putative virus species. This was because many of these newly discovered virus species were confined to only one of the newly discovered *Pelagibacter* species.

This strong infection preference was apparent even though several other closely related *Pelagibacter* species were present in the same samples, and would therefore have been available to be infected. It is the pattern we would expect if different virus species have evolved to target different host bacteria.

When G.E. Hutchinson called attention to the apparent paradox of the plankton back in 1963, he was primarily concerned with the problem of how such a seemingly simple environment can maintain so many species. The species that he knew about when he wrote his paper were mostly tiny plant and animal species that could be seen with the naked eye or under an ordinary microscope. The *Tara* Oceans workers had discovered an even greater diversity among the even tinier bacterial and virus species.

Some ecologists who followed Hutchinson proposed that the answer to his paradox was simply that the physical environment in open oceans and lakes must be more variable than it first appears. Such a diverse environment should provide ecological niches for many species. And they are partly right. Variations in temperature, water clarity, depth, and patterns of currents do indeed provide physical diversity. And the bacterial and virus strains in the *Tara* study are indeed non-randomly distributed across temperatures and depths.

But single-cell genomics is now showing that the physical characteristics of the ocean contribute only a small part to ecosystem evolution. Dominating such evolution, as Darwin proposed for the species in the Galápagos, are the entangled relationships among species. These entanglements provide proliferating opportunities for the evolution of, in Darwin's immortal phrase, the "endless forms most beautiful and most wonderful" that can take advantage of them.

The Perils of Tiny Organisms

As is obvious from daily news reports, the goings-on in the microbial world are not all benign. We must exercise great care if we attempt to manipulate microbiomes, because disturbances of these bubbling evolutionary cauldrons can yield unexpected results.

The disastrous introduction of plague and leprosy into Europe, and of smallpox, measles, and influenza into the New World, along with the more recent rapid spreads of zoonotic diseases worldwide because of our new modes of travel, demonstrate some of the dangers to our own species. Many diseases that affect wild and domesticated animals and birds, in addition to diseases that infect the plant species that we all depend on, can also have devastating effects.

Cholera provides a vivid example of these lurking dangers. Cholera is caused by a highly motile bacterium, *Vibrio cholerae*. It is one of many species in the *Vibrio* genus, twelve of which are known to cause diseases in humans.

V. cholerae can produce severe and life-threatening diarrhea and uncontrolled vomiting by permanently switching on transport proteins in the membranes of the victim's gut cells and causing the gut tissues to lose massive amounts of water. But not all strains of the bacterium can do this. Mobile elements called pathogenicity islands, which can easily spread both within and between bacterial species, carry the genes necessary for the pathogenic effects. And bacteriophages can also carry pathogenicity genes from one *Vibrio* to another and even between *Vibrio* species.

Rita Colwell of the University of Maryland and her many colleagues have carried out detailed studies of where *Vibrio* strains and species live. They find that these bacteria are widespread in marine environments [74]. *Vibrio* normally infect tiny zooplankton and crustaceans, and play roles (as so many planktonic bacteria do) in the overall economy of these ecosystems. In particular, they help to control crustacean populations and recycle their nutrients. And they can easily jump from those environments or from sewage to humans.

Because *Vibrio* can gain or lose pathogenicity genes, quickly switching from more to less virulent strains and back, the degree of their virulence depends on the density of hosts. The bacteria will tend to be selected for high virulence if many hosts are available. Under these conditions, the bacteria can spread easily—even though they may kill their hosts in the process.

The bacteria react only to present conditions and cannot anticipate the future. They will be selected for characteristics that allow them to do whatever it takes to have lots of offspring. They will be selected to kill their hosts if the effect of killing them is that their own rate of spread to other hosts is increased. They will be selected to *stop* killing those hosts if such extreme behavior leads to fewer hosts and a decrease in the rate of spread. What is unusual are the number of ways by which these evolutionary changes can be accomplished, thanks to the ubiquitous presence of pathogenicity islands and pathogenicity-altering viruses in these bacterial populations and in their immediate environment.

Some of the other *Vibrio* species have the potential to pose even greater risks than *V. cholerae*. Dr. Colwell has been investigating rare infections of swimmers in Chesapeake Bay by *Vibrio vulnificus*. This bacterium, like *V. cholerae*, normally preys on and checks populations of marine crustaceans. Its infections in humans can spread swiftly from even small open wounds. They have a 50 percent mortality rate, and the survivors often lose limbs [75].

Dr. Colwell told me that whenever she gives a talk about this pathogen, there are always one or two people in the audience who have lost a limb to it. Her lab and other labs have shown that this horrifying bacterium is becoming more common in the bay as its waters heat up from global warming. These changes mirror the large recent increases in *V. cholerae* populations that have been seen in the Atlantic basin and that are possibly connected to increases in water temperature [76].

Extensive surveys of gut metagenomes from healthy humans from around the world have turned up the occasional presence of harmless variants of three different *Vibrio* species, including *V. cholerae* [77]. Could these minor components of our microbiome be potential time bombs, ready to be ignited by the right environmental circumstances?

Vibrio cholerae is constantly in a state of evolutionary flux because of its many interactions with a range of host species and gene transfer agents that span more than one ecosystem. Genetic changes can quickly make the bacterium dangerous to humans, and can just as quickly reduce that danger. Cholera is extremely dangerous, but we are lucky that reductions as well as increases in virulence can happen so readily.

This is not the case with the smallpox virus, a large and complex DNA orthopoxvirus. This virus has caused severe disease in humans for millennia, and its toll has been immense. It is estimated that during the period 1880–1980, it killed half a billion people. In 1980, populations of the virus were finally eradicated, in one of the most transformative public health achievements of the twentieth century.

The origins of this virus are mysterious. Other orthopoxviruses can infect several mammalian species, but the smallpox virus seems to have had no host other than humans. Perhaps it originally spread to our species from one of our now-extinct primate relatives.

Smallpox was also unusual because it could not survive at human body temperature. The only way it could spread was by infecting the skin, which is several degrees cooler than the rest of the body. The oozing, life-threatening, skin-destroying rash that the virus triggered facilitated its transmission.

Smallpox, unlike *V. cholerae*, seems to have been insulated from the genetic changes that are continually being generated in the evolutionary cauldrons that surround us. The smallpox virus succumbed to our eradication efforts not because it changed, but because it didn't change. Safe and successful vaccination against this disease, using its less pathogenic cowpox virus relative, was introduced in the West at the end of the eighteenth century. But crude and dangerous vaccinations with small amounts of live smallpox virus had been carried out in China since at least the tenth century, and the practice had spread widely. The virus, perhaps because it was cut off from many sources of genetic change through recombination and gene transfer from other viruses, stayed highly susceptible to the antibodies induced by vaccination for the entire 1,000 years between those early days of vaccination in China and its final eradication.

We were lucky that a combination of smallpox's genetic isolation and advances in immunology research allowed us to conquer this agent of death. But we have been warned.

Imagine that, in the future, we deliberately or inadvertently change microbiomes on a massive scale so that they start to lose their ecosystem-stabilizing diversity. An accidental introduction of microbes with the potential to become dangerous pathogens into these simplified ecosystems could prove disastrous. And if, through such damage to the world's ecosystems, we deplete the possibilities for genetic change in future pathogens, we will only have ourselves to blame for the uncontrolled spread of diseases. Such risks are present even if we do not consciously manipulate ecosystems, as the increase of *V. vulnificus* in Chesapeake Bay shows.

But there are good ways of lowering these risks, which we will explore in the book's final chapter. They depend on keeping those evolutionary cauldrons bubbling.

6

Swift Evolution in Tiny Entangled Banks

Entanglement: something that entangles, confuses or ensnares.
Merriam-Webster

Earlier, I promised to tell some tales of how laboratory experiments can provide a path to understanding how, and how quickly, the evolution of interactions between organisms can produce an entangled bank. Now that we have the needed background, we are ready to explore some remarkable recent experiments that provide answers to these questions. The stories are rather complicated, but the results are amazing.

Even the Simplest Ecosystems Can Rapidly Give Rise to Complexity

In the early 2000s, bioengineer Norikazu Ichihashi and his colleagues at Japan's Osaka University began to speculate about the role that the evolution of early ecosystems might have played in the evolution of life itself. They wondered whether it might be possible to start with a truly simple ecosystem, perhaps of a kind that could have been present near the dawn of life, and then follow exactly how it changes over time.

Ichihashi's group was primarily interested in the evolutionary and ecological paths that early life on the Earth might have followed. But in the process of analyzing their results, they also cast a remarkable new light on the evolution of ecosystems in general.

They began by asking if a really, really simple ecosystem can evolve to become more complicated and support more inhabitants—and if so, how quickly this could happen. And what would they find if they used the powerful tools of molecular biology to follow the details of exactly what was going on as the ecosystem began to evolve?

Ichihashi knew of earlier work on a virus called Q-beta, which preys on bacteria. This virus has an honored place in evolutionary studies. During the 1970s, Sol Spiegelman of Columbia University had used parts of the virus to create a simple evolving system inside a test tube.

Spiegelman selected for rapid replication of the chromosome of the virus. This virus chromosome is a molecule of RNA rather than of DNA. The Q-beta virus can use its RNA-dependent RNA polymerase to make copies of its chromosome, so that the virus does not need DNA at all.

By cutting DNA out of the picture, the virus has cut down on the number of things it must do to duplicate itself. RNA is more fragile than DNA, and it mutates more easily. This drawback is balanced by the nature of the virus populations themselves. If such a population accumulates many mutations, this may actually be more of an advantage than a disadvantage. Even the occasional rare mutant virus strains with high fitness can quickly grow to vast populations.

This ability to overcome a high mutation rate may benefit both the viruses that can slaughter one third of their host bacteria every day and their much-abused hosts that must fight these homicidal (germicidal?) viruses to a draw. Theses daily battles may actually put a premium on both a high virus and a high host mutation rate.

In his experiments, Spiegelman allowed the high mutation rate of the virus to operate almost unchecked. He purified the enzyme that the virus used to replicate its RNA chromosome, froze it, and used samples of this purified *replicase* enzyme throughout his experiment. To initiate the experiment, he added RNA building blocks to a test tube containing the replicase and then topped the mix off with copies of the virus RNA chromosome. After the replicase made copies of the chromosome, he transferred some of these copies to a fresh solution of replicase enzyme and RNA building blocks, in order to repeat the process and make more copies of the virus chromosome [78]. He kept doing this for dozens of "generations."

This was not a living system, since at no time were host bacterial cells or complete viruses involved. The replicase's only task was to make more copies of the virus chromosomes that it was given.

In the real world, if the replicase were to make a defective virus chromosome that lacked an essential function, that chromosome would immediately be lost from the population of viruses. But in Spiegelman's experiments the chromosomes that tended to dominate the virus chromosome population after repeated transfers were simply the ones that could be replicated swiftly by the enzyme, regardless of the chromosomes' other properties. The only requirement was that any mutations should not destroy the signals on the virus chromosome that were needed for the replicase to bind to it and begin to replicate it.

The Q-beta chromosome is a molecule with a length of 4,500 nucleotides, but the copies of it that survived repeated transfers rapidly became shorter and shorter because of selection for ever-swifter replication.

Eventually Spiegelman hit a wall. The replicase was unable to replicate a molecule that was shorter than 288 nucleotides. This much-shortened molecule, even though it had lost most of its genes and the ability to invade bacteria, could still multiply with blinding speed in Spiegelman's cell-free system. This stripped-down remnant of a virus chromosome became known as "Spiegelman's monster."

Of course, unlike laboratory-created monsters in science fiction movies, this monster was utterly incapable of surviving anywhere but in Spiegelman's simplified world. The monster was able to tell Spiegelman much about the information such a stripped-down RNA chromosome still needed in order to be replicated properly by the replicase enzyme. And the fact that the monster had lost so much compared to its parent chromosome also showed that in the real world, there are hundreds or thousands of selective pressures that must act in order to keep the virus chromosome at its original full length and to maintain its many functions. The absence of those pressures in the Spiegelman experiment produced a molecule that was really, really good at just one thing, rather than one that was good at hundreds or thousands of things.

Ichihashi took these experiments much further. He asked whether this simple system could be modified to allow more than one selective pressure to operate, and if so what the result might be.

When the Q-beta bacteriophage is floating around in the environment between bacterial hosts, it cannot employ its polymerase to make copies of itself. The free-floating virus simply exists in a kind of limbo. But when it invades its host, the common gut bacterium *Escherichia coli*, that host cell provides everything that the virus' RNA polymerase gene, and that its other genes, need to make a new generation of viruses.

NASA engineers who build a Mars probe must strip it of all nonessentials to keep its weight down. Like these engineers, evolution has removed from the virus' chromosome everything except what it absolutely needs in order to invade its bacterial hosts and replicate there.

Indeed, evolution has gone further. The virus polymerase gene carries only part of the information that its protein needs to function as a working RNA polymerase molecule. The virus protein must also be able to bind to and co-opt three additional proteins that are borrowed from the bacterial host. These proteins normally contribute to the bacterium's DNA replication machinery, but now they help to spell the bacterium's death sentence.

When augmented in this way, the replicase molecule is able to use the bacterium's store of RNA building-block molecules to make many copies of the virus' RNA chromosome. These copies of the virus chromosome can cannibalize the entire bacterial protein synthesis machinery, feeding their genetic information into it to synthesize the other virus proteins that are needed to form complete virus particles. The new virus chromosomes are then enclosed by the new virus protein particles.

This new generation of viruses, like the spawn of the creatures in an *Alien* movie, then bursts through the bacterial cell walls and floats away. Most of these new viruses will be lost, but a few will be lucky and encounter other (unlucky) host bacteria.

This hybrid virus-bacterium RNA polymerase is the same as the one that Spiegelman used to make more copies of the virus chromosome. During Spiegelman's experiments, the polymerase was always taken from the stock in his freezer and remained unchanged throughout. But suppose the polymerase itself was free to mutate and evolve?

Ichihashi realized that the virus chromosome, instead of simply being a passive accumulator of mostly chromosome-shortening mutations, could play a more active role in a selection experiment. The virus' replicase gene, and therefore the replicase protein that it coded for, could be forced to take part in the evolutionary process.

He began by cloning the part of the virus chromosome that carries only the gene for its RNA polymerase molecule. This piece of RNA can be used by the host's cellular machinery to make a functioning polymerase. But to do this, the RNA fragment needs a lot of help.

Each generation of the experiment, Ichihashi and his colleagues provided this help in the form of a molecular cocktail. It consisted of the building block molecules that were needed to make RNA, the bacterial proteins that helped the polymerase to do its job, and a complete kit of the bacterial host's protein-making machinery that he purified from the cells of a virus-free bacterium host. It was this cocktail, rather than the virus chromosome, that remained unchanged throughout the experiment.

With the assistance of the cocktail, the virus polymerase gene that he had cloned could make the virus part of the polymerase protein molecule. The cocktail itself supplied the rest of what was needed to make a functioning polymerase, and this new polymerase could then make new copies of the virus polymerase gene.

When those new copies of the polymerase gene were transferred to a new container with fresh bits of protein-making machinery, the copies that were fully functional could make more polymerase. So could copies with mutations that had a slight effect on the polymerase. And copies that no longer functioned but were still able to be replicated could survive long enough to be replicated by the polymerase. Thus, many kinds of mutant polymerase genes could survive in this experiment.

The Ichihashi lab made suspensions of tiny droplets in oil, containing this mix of virus and host molecules. The RNA and protein synthesis reactions proceeded as far as they could in each droplet, halting when they ran out of building blocks.

Periodically, the experimenters added new droplets to the suspension. These droplets did not contain more virus RNA, but they did contain all the bacterium host components that were needed for further protein and RNA replication.

This system had only a tiny fraction of the components of the original virus. But what it could still do was pretty amazing. It consisted of a single gene coding for a protein that, with the help of the cocktail, could make new copies of the gene. These new copies of the gene were then given the resources needed to make the protein, and this process could be repeated over and over for many generations.

Because most of the components were provided anew each generation and did not change over time, the only thing that could evolve over time was the coded RNA information in the virus gene. In Spiegelman's experiment, the most successful mutations were those that shortened the chromosome so that it could be replicated more quickly. But in Ichihashi's experiment, a new universe of possibilities opened up. The mutational changes in the Ichihashi experiment could alter the virus polymerase protein in many ways, and these changes would have a chance to survive as long as the mutant virus replicase retained its ability to replicate RNA.

This system was far more demanding of its pieces of RNA than Spiegelman's original experiment. At least some of the polymerase genes in the system had to retain the ability to code for a functioning RNA replicase protein, or the whole system would stop. But that left room for a zoo of possibilities. Because of all the mutations that were now possible, the original simple system quickly began to generate complexities. These different sequences warred with each other for the resources of the cocktail, and were continually faced with new challenges as they were mixed and shuffled.

Within the first few generations of replication of Ichihashi's small RNA, mutant RNAs had begun to appear in the droplets. Some of the changes were tiny, only slightly altering the function of the polymerase molecule that they coded for. Other, more dramatic, mutant molecules resembled those that Spiegelman had originally selected for. These mutants had lost part of their RNA, so that they no longer coded for functioning replicases (Figure 6.1). Some of these crippled mutants could, however, still be replicated by a functioning replicase enzyme.

Each generation, Ichihashi broke the cocktail of molecules into little droplets, and used ingenious methods to mix adjacent droplets together at the end of each generation. Each droplet therefore had finite resources during the replication process, but was then mixed with other droplets and provided with fresh resources. This greatly increased the intensity of selection within each tiny droplet.

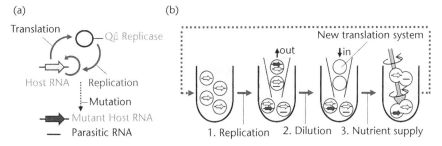

Figure 6.1 Ichihashi's replicating system in diagram form. The RNAs that are still able to be translated into polymerase are shown with small arrows, gray for the original RNA and black for a mutant but still-functional RNA. Mutant non-functional RNAs are shown as simple short lines.

In each droplet, as long as copies of these shortened RNAs could still be made by other replicase molecules that had retained their function, those copies could survive and perhaps be replicated in their turn during subsequent generations.

These shortened or otherwise defective RNAs could not take over the population, because they no longer coded for the functioning replicase molecules that were needed to replicate them. But some of them could become *parasites* on the system. Old copies of the defective RNAs would be diluted away, but new copies could be produced each generation. These molecules were free riders that co-opted some of the resources in the droplet. And sometimes these parasitic copies could persist for many generations, even though they wasted the time of the functioning polymerases and caused them to use up precious resources to make meaningless copies of these meaningless mutant RNAs.

If I may mix a metaphor, these parasites had to walk a narrow line. Only the copies of parasitic RNAs that were still large and slowly replicating, so that they did not use up too many of the droplets' resources, could persist. Nonfunctional RNAs that were really good at being replicated—perhaps because they were shortened like Spiegelman's monster—would gobble up all the available resources. They would be lost, and they would likely take the rest of the molecules in their droplet with them.

It was the parasitic RNAs that were able to walk the line successfully that lasted for many generations. Just as in a population of living organisms, natural selection acting on this laboratory system could remove the most harmful mutations while retaining some less-harmful ones.

The result was a diversifying zoo of molecules, including parasites, that were evolving before the scientists' eyes. By sequencing the virus polymerase genes periodically, the scientists could determine the exact mix of sequences that were present. They could determine the precise nature of the mutations that had happened, exactly when they had happened, and how far they had spread in the population. And the scientists could measure what the different mutant polymerase molecules were able to do, by purifying each of them and comparing their abilities to copy the various RNAs that were in the mix.

Selection in this system was fierce, and did not just act to weed out defective mutant polymerases that were too "greedy" and used up too many resources. Even functioning molecules that were a bit slow off the mark were severely penalized.

Within a short time, this collection of intensely competing RNA and polymerase molecules had evolved to become a kind of mini-entangled bank [79].

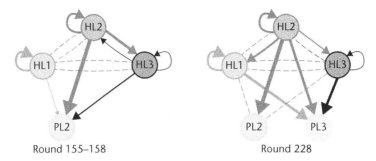

Round 155–158   Round 228

Figure 6.2 Left: the replication capabilities of the four major RNA molecules at generation 155. Right: the capabilities of the five major molecules, including two parasitic forms, at generation 228 when the experiment was stopped. The thickness of the arrows shows how successfully the functional RNAs could replicate themselves and each of the other RNAs in the system.

The polymerase genes had branched into different "species," which interacted with each other in a variety of ways.

Ichihashi's group measured the interactions among the commonest of these molecules by setting up experiments using different combinations of polymerases and RNAs. Arrows in Figure 6.2 show some of the interactions and how they were continuing to evolve.

By generation 155, no original polymerase genes were left. The original molecules had been replaced by three slightly different functional polymerase genes, all of which were coexisting in the population. Ichihashi termed these functional lineages Host Lineages 1, 2, and 3, because they acted as hosts for parasitic molecules. These RNAs could use the supplied cellular machinery to make polymerases that were able to copy parasite as well as functional RNA.

All of these host lineages were able to code for replicases that could replicate themselves and other genes, but they did so with varying degrees of success. In addition, there was a parasite lineage, designated PL2. (A parasite named PL1 had appeared early on, but had vanished.)

By this time, a little over halfway through the experiment, the interactions among these molecules had already become complicated. HL3 could barely replicate itself, but it was readily replicated by HL2. HL1 could barely replicate the parasite lineage, but the other two host lineages could do so easily.

By replication round 228, when the experiment was terminated, the surviving descendants of the original replicase RNA had accumulated further mutations.

The host lineages had changed little, so the experimenters retained their designations. But those small additional mutational changes had altered the properties of the host lineages slightly. HL2 had gained some ability to replicate HL1, but lost some ability to replicate HL3. And a new parasite lineage, PL3, had appeared. Both HL1 and HL3 had lost the ability to replicate PL2, but they were easily able to replicate the new PL3.

The host replicases were obviously a mixed bag, with HL2 being much more talented and versatile than the others. The parasite RNAs differed in their ability to sponge off the system. Yet all of these different molecules were managing to survive through many rounds of replication.

Much rarer RNAs and replicases abounded, but they came and went rapidly. Most were probably unsuccessful mutants.

And there were hints of further complexity. The new lineages of the host molecules had apparently acquired new properties. The experimenters found that some of them appeared to aid each other. They could replicate more quickly in mixtures than when they were by themselves. And, puzzlingly, the HL3 replicase was found to be unable to replicate its own RNA when it was by itself, suggesting that some unknown interaction with other molecules in the intact ecosystem was enabling HL3 to persist.

Work on this exciting ecosystem continues. Ichihashi's group has recently shown that systems of mixed host and parasite sequences are very likely to arise in such a system, with a parasite arising first followed by some copies of a new host that is resistant to replicating the parasite [80].

Even in such a simple system, hosts and their parasites can quickly generate new ecological niches. Darwin had suspected that for most entangled banks it was interactions among species that were the primary drivers of evolutionary change. The work of Ichihashi and his group, in addition to showing that complex evolutionary entanglements can emerge with surprising swiftness, suggests that such forces may have been operating even at the time that life was just emerging.

In the Introduction to this book, we explored part of an entangled bank of interactions in the Amazon rainforest. We have since encountered many such entangled banks. But Ichihashi's experiments show vividly how even a highly simplified approximation of a living ecosystem is sufficient to produce some entanglements. They also show how some of these molecules, even when damaged, could still game the system. These parasitic sequences may have been mindless, but like George Washington Plunkitt they were still capable of "seein' their opportunities an' takin' 'em!"

In 1976, in his seminal book *The Selfish Gene*, Richard Dawkins suggested that genes that do nothing beyond replicating themselves should evolve readily, an idea that has been proved correct multiple times since. Ichihashi's experiment shows how quickly and easily this can happen. But parasitism in the real world is usually far more complicated than simply the ability to sponge off the system. It is now becoming clear that many evolutionary entanglements have evolved and spread through interactions between host organisms and their sometimes extremely complex and versatile parasites.

7

The Boundless Potential of
Evolutionary Entanglements

…There is something about a Martini,
Ere the dining and dancing begin,
And to tell you the truth,
It is not the vermouth—
I think that perhaps it's the gin.
Ogden Nash, "A Drink With Something In It"

An Ecosystem with Something in It

In the 1970s, before the introduction of DNA sequencing and before the invention of most of the many currently available ways to clone and manipulate genes, my laboratory was working on the bakers' and brewers' yeast *Saccharomyces cerevisiae*. These tiny one-celled organisms are responsible for the production of ethyl alcohol and for the CO_2 gas bubbles that form in rising dough. It is the source of three of our species' happiest inventions: croissants, wine, and martinis.

The enzyme that we were working on, which plays a central role in those wondrous inventions, is the yeast version of the enzyme alcohol dehydrogenase. When we began our work, a huge controversy was dividing evolutionary biologists. Japanese population geneticist Motoo Kimura had proposed that most of the evolutionary change taking place at the DNA level must be selectively neutral [81]. He reasoned that genetic variants that actually bring about fitness differences should be rare. This is because, if large numbers of these differences are acted on by natural selection, this would place such a large selective burden on populations that the populations themselves might be driven extinct.

Like many others, my group was unconvinced by his reasoning. I suggested a "selectionist" answer [82] to the conundrum posed by Kimura. I pointed out that many of the selected variants that spread in populations are *selected functional polymorphisms*, which are able to do their job more effectively than the variants that they replace. But the variants that are replaced are still able to function during the replacement process, so the burden on the population is not as great as it would be if they were no longer able to do their job. If this type of polymorphisms were common, many selected variants should arise in populations without causing a crushing burden.

We looked to see whether we could produce such new functional genetic variants by artificial selection in yeast populations. We planned to examine their properties in detail and follow their fate in these populations. How would these new mutants behave? Would the alleles they were replacing still be able to function? Would these new mutations impose a burden on their carriers, and if so, how much of a burden?

We soon found a mutant variant of yeast alcohol dehydrogenase, the enzyme responsible for ethyl alcohol production. The variant permitted the cells carrying it to survive a severe environmental challenge that was detrimental to normal yeast cells.

This enzyme can work in both directions, turning the important metabolite acetaldehyde to ethyl alcohol and back again. We challenged the cells by introducing allyl alcohol, an alcohol related to ethyl alcohol, into their growth medium. Yeast's alcohol dehydrogenase can cleave this alcohol, but the result is a poisonous aldehyde.

We found that our mutant enzyme could still play its normal role in the cell's metabolism, while producing the nasty aldehyde at lower levels than the yeast's original enzyme did. This allowed the mutant cells to take over the population.

Hans Jörnvall of Sweden's Karolinska Institute and I tracked down the amino acid change that was responsible for the mutation, a substantial task in the days before DNA sequencing [83]. The mutation turned out to involve a single amino acid near the enzyme's reactive site, a replacement of the amino acid histidine with the chemically different arginine. This change slowed the rate at which the mutant enzyme carried out its reactions, and also made it less likely to cleave allyl alcohol.

In subsequent work, my lab found that this small change in the enzyme had a profound effect on the chemistry of the entire cell [84]. The small mutational change that we had selected for in the molecule was actually capable of changing the balance between basic and acidic forms of a cofactor molecule used by alcohol dehydrogenase and many other enzymes. It was this shift that made the mutant enzyme much less likely than the original enzyme to act on allyl alcohol and produce the poisonous aldehyde. In spite of this dramatic change, the mutant cells were still able to function.

This mutation was a selected functional polymorphism. If small amounts of allyl alcohol were introduced into the cells' environment, the mutant allele did not place a huge burden on the population as it spread. Instead, the mutant cells survived well under conditions in which the original non-mutant cells grew relatively poorly. This provided an enormous positive boost to the mutant cells' fitness and still allowed non-mutant cells to survive.

Now, decades later, experiments can be done to explore gene pools in unprecedented detail, using whole-genome analysis. These newer studies show that selected mutational changes do indeed happen much oftener than Kimura would have predicted. Neutral changes can also appear, playing a less signifi-cant but still potentially important role. These experiments are also able to probe the capacity of entire gene pools to evolve and meet new challenges.

An especially illuminating recent study, by Australian Michael McDonald and his co-workers [85], took advantage of whole genome analysis to follow some of the ways by which a single gene pool can change with time.

They started with a strain of *Saccharomyces cerevisiae* yeast, which could be grown up as genetically uniform copies of a single originating cell. They repeat-edly transferred cells of these strains to fresh growth medium. Under these con-ditions, and given plentiful food, the yeast cells simply budded off exact copies of themselves, just as bacteria do most of the time.

But yeast cells, like our own cells and unlike bacteria, have true nuclei. When the yeast cells are grown on a nitrogen-poor medium, they switch on exactly the same gene-shuffling and gamete-producing mechanisms that we use during our own sexual reproduction.

In the McDonald experiment, half of the yeast strains were chosen to undergo periodic rounds of such sexual recombination. In the rest of the strains, the cells were simply allowed to multiply asexually.

Periodically throughout the experiment, DNA was isolated from both the sexual and asexual populations and massively sequenced using the same methods as metagenome analysis. But in this case, it was the many cells of a single yeast population that were being sequenced.

Each of these mass sequencings yielded a set of overlapping fragments. When the fragments were pieced together, they gave the McDonald group detailed views of that strain's entire gene pool.

New mutations soon began to appear. They were easy to detect because they were taking place in a population of yeast cells that had originally all shared exactly the same set of genes. Any mutation that happened anywhere on the yeasts' chromosomes could be detected by this "shotgun" sequencing, so long as it had reached a frequency of 1 percent or so.

McDonald and his co-workers undoubtedly missed many mutations that were neutral or harmful in their effects and that quickly disappeared. But they were able to track the numerous other mutations that became common enough to be detected. Some of these mutations swept through the entire population, others rose to intermediate frequencies, and many fluctuated in frequency over time. Because the entire yeast genome sequence was known, it was possible to identify with absolute certainty the exact locations of all these mutations on the yeast's chromosomes, and the exact changes to the DNA that were involved in each mutation.

Because some amino acids are specified by more than one code word of the genetic code, some of the changes that they saw in the coding parts of the DNA were "silent" or *synonymous*, like many of Marty Kreitman's mutations. Even though these mutations changed the DNA, they did not change the protein coded by that gene. Such synonymous changes might be expected to fall into Kimura's selectively neutral category.

Other changes were *non-synonymous*, which meant that they did alter the amino acid sequence of the proteins. These mutations were more likely to have positive or negative selective effects.

In addition, some of the changes happened in *intergenic* parts of the chromosome that do not code for genes. Such changes have been assumed to have the same effect as synonymous changes in the coding regions. This may not be the case, especially in complicated organisms, but in the McDonald experiments, they did indeed behave as if they were selectively neutral.

Figure 7.1 shows that many of the mutations that appeared in the asexual populations subsequently swept through those populations and were "fixed"— that is, they replaced, or nearly replaced, the older alleles that the strain had started with. And, remarkably, the proportions and even the numbers of these mutations that fell into the synonymous, non-synonymous, and intergenic categories hardly changed during the fixation process.

Kimura would not have predicted these results. The overwhelming majority of these different kinds of mutations, which must surely have included some with negative effects, were sweeping through these populations as if they were being selected for, while very few were being lost. This should not have

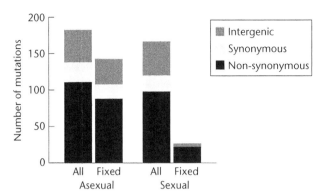

Figure 7.1 Bar graphs 1 and 3 show that intergenic, synonymous, and non-synonymous mutations all appeared at comparable rates in the asexual and sexual populations. Bars 2 and 4 show that genes that swept through the asexual and sexual populations differed greatly in the asexual and sexual populations. Many mutations of all types swept through the asexual populations. Far fewer swept through the sexual populations and most of those that did were non-synonymous genetic changes.

happened. Most mutations, including most neutral ones, should disappear from populations shortly after they appear.

Luckily, the very different results that McDonald's group obtained from their sexually reproducing populations explain what must have happened. In these populations, as would be expected because of the random nature of mutations, the initial proportions of synonymous, non-synonymous, and intergenic mutations were not significantly different from those that arose in the asexual populations. But far fewer mutations were "fixed" in the sexual populations. Almost all of those that were fixed were non-synonymous, presumably because they had a positive functional effect on their proteins.

In the asexual populations, the genes in each cell are inherited as a unit and almost never recombine. This means that each chromosome acts as a whole block of genes. So, if an especially favorable mutation were to happen in that cell, it could drag along with it any and all other mutations that had appeared anywhere in the same lineage. These other mutations would be hitchhikers.

In the asexual populations, both synonymous and intergenic mutations were swept (or dragged) to fixation in high numbers. Most of these mutations were presumably neutral or nearly so, and could essentially hitchhike for free. But, if the positive selection were strong enough, the advantageous mutation might have been able to drag slightly harmful mutations with it. What is especially surprising about McDonald's data is how often this seems to have happened in the asexual populations. Almost all of the non-synonymous mutations that had been detected in the asexual populations were also fixed, even though many of them must have been at least to some degree harmful because they made random alterations in their proteins.

It seems possible that all this puzzling fixation in the asexual populations can be traced to a relatively few super-fit mutations. So fit are these super-mutations that, once they appear, they can drag along with them to fixation the entire menagerie of other mutations in the cell's genome. These super-fit mutations are like the kind-hearted driver of a big truck who encounters a heterogeneous crowd of hitchhikers on the road, gathers them all into the back of her truck, and sweeps them to their destination.

This interpretation of the results is reinforced by the data from the sexually reproducing populations. Sexual reproduction results in the breakup of the genes on a chromosome through genetic recombination. In these populations, it appears that the linkages between advantageous and harmful or neutral mutations are indeed largely broken up, so that fewer of these neutral or harmful genes are dragged along by hitchhiking. All the synonymous mutations and most of the intergenic ones either are lost or remain at low frequency in these populations. And, of the intergenic mutations, only a minority are fixed. I suspect that some of these mutations in the sexual population must be the equivalent of super-fit mutations in the asexual population, but such an equivalence will be challenging to demonstrate.

It is fascinating that, in the sexual populations, many non-synonymous mutations are lost but a substantial number have become fixed. This suggests that, despite repeated predictions by population geneticists that advantageous mutations ought to be rare, such mutations can appear at high rates. And remember that those yeast populations were growing without competitors and under uniform conditions. Even this unexciting environment seems to have provided ample opportunity for the appearance of advantageous new mutations, along with what are probably many new balanced polymorphisms in which opposing selective pressures keep alleles at intermediate frequencies.

The effect on the overall fitnesses of the populations was consistent and dramatic. Figure 7.2 shows the gain in average competitive ability of the sexual and asexual populations when they were matched against the original strain with which the experiment started. The asexual strains showed a significant gain in relative fitness against the original strain, but the sexual strains gained twice as much, an astonishing 15 percent on average. This increase in fitness, again contradicting Kimura, did not impose a crushing burden on the population.

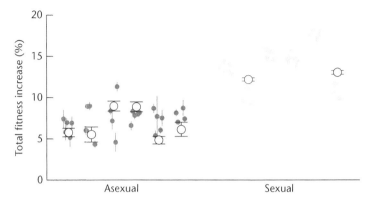

Figure 7.2 Difference in fitness between asexual and asexual strains at the end of the experiment, compared against the original strain used to start the experiment.

These remarkable results suggest that sexual recombination can purge populations of most harmful and even neutral mutations, so that most of the surviving mutations are advantageous. McDonald and his co-workers had found strong evidence that sexual recombination is indeed advantageous.

Subsequent experiments by the McDonald lab yielded a more nuanced picture [86]. As more variables were added, the results became more complex. In particular, allowing the cells to compete as diploids, in which each cell carries one set of maternal and one set of paternal chromosomes, along with the addition of greater variation in the experimental environment, seemed to provide more, and more complex, paths toward higher fitness. Perhaps McDonald was seeing a complexification of his yeast ecosystems similar to what Ichihashi saw in his stripped-down little ecosystem. But the overall pattern, in which sexual reproduction breaks up hitchhiking groups of genes and provides a powerful way to sort out and recombine advantageous variants into new combinations, was preserved in those later experiments.

These experiments make clear why sexual reproduction represents such a huge advantage. It can scramble gene pools, purge them of harmful mutations, and provide new evolutionary opportunities. The experiments make plain why most species of organisms are able to recombine their genes.

McDonald's work also showed that super-advantageous genes may not, contrary to the expectation of most population geneticists, be as rare as unicorns. Because such super-fit mutations seem to be moderately common in yeast, they will probably turn out to be moderately common throughout the living world.

Genetic Transfers that Draw on the Entire Tree of Life

McDonald's yeast experiments suggest that super-advantageous mutations can appear in a population even when there are no competing species present. But in the real world and in every ecosystem, even in the apparently featureless real world of the open ocean, a seething mass of competing species creates an evolutionary cauldron at the microorganism level. Such churning cauldrons must provide entirely new opportunities for super-advantageous mutations and other kinds of genetic change that can draw on the power of genetic recombination and other mechanisms of gene transfer. Laboratories around the world are starting to explore these possibilities, which are turning up everywhere.

Studies are revealing that, in addition to frequent pseudo-sexual events, infrequent horizontal gene transfers that can introduce new genes into distantly related species are also playing a more important part in the evolution of ecosystems than we could have imagined a few years ago.

In 2011, Chris Smillie, Eric Alm, and their co-workers at MIT [87] examined more than 2,000 complete bacterium genomes that had been pieced together from metagenomic analyses. The microbial communities that they analyzed were mostly microbiomes of the human gut, mouth, and skin. The sample also included microbiomes from domestic animals and from various types of soil.

Clear evidence that horizontal gene transfers had taken place was provided by scans of the genomes of bacteria in the microbiomes for sequences of bases that they shared with other bacteria living in the same microbiome. If two different bacterial species in a microbiome carried identical sequences of DNA that were more than 500 bases long, and especially if the two species were distantly related, this was taken as a sign that recent horizontal genetic transfers between the species had occurred. Older transfers between the species had also probably happened, but they would have accumulated many different mutations in the two descendant lineages, and would no longer have been identical.

To reduce the likelihood that these matching sequences were not simply identical because by chance they had not undergone mutational change, the comparisons were limited to pairs of species whose slowly evolving 16S RNA genes were at least 3 percent different from each other. This restriction ignored any evolutionary separations between species of less than 150 million years.

Figure 7.3 shows the number of signals of horizontal gene transfer that could be detected in these microbiome entangled banks.

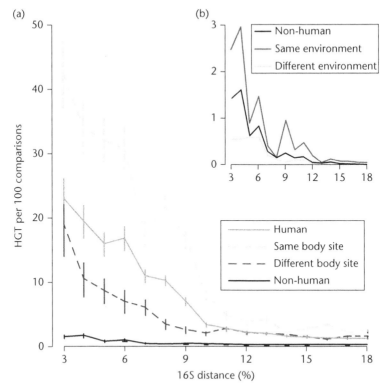

Figure 7.3 Numbers of horizontal gene transfer (HGT) events between bacterial species inhabiting the same microbiome. The various bacterial species are separated by a wide range of evolutionary distances, as measured by the percent difference between their slowly evolving 16S ribosomal RNA genes.

The vertical axis in the figure marks the number of such long shared pieces that were found in every 100 comparisons between two bacterial genomes taken from that microbiome. The horizontal axis marks the evolutionary distance between the bacterial species that were being compared. The species comparisons range all the way from those between relatively recently evolved pairs of species (small 16S differences) to those between pairs of species that are separated by billions of years of evolution (large 16S differences).

The graphs are rich in information. Graph **a** summarizes all the data. Clearly, horizontal gene transfers have had their greatest impact on the closely related species of bacteria that share the same human microbiome—gut, skin, or mouth. These high numbers of transfers decline precipitously with increasing evolutionary distance between the bacterial species. Even among the most distantly related bacteria, however, their numbers do not drop to zero.

There are also huge differences in the numbers of transfers between bacteria that inhabit the same body site and those that inhabit different sites—the gut and the mouth, for example. It appears that a tiny physical distance has been enough to produce big reductions in the number of horizontal gene transfers between them. In human microbiomes, such seemingly tiny distances as the distance between the gut and the mouth have had a big impact on the evolutionary histories of their small inhabitants.

Graph **b**, enlarged from the bottom line of graph **a**, shows that horizontal transfers have played a far smaller role in microbiomes of non-human animals and of soil samples. But the overall shapes of the curves resemble the curves from humans. Both these non-human types of microbiome also show declines in numbers of transfers as the evolutionary distance between the bacterial species that are being compared increases.

Human microbiomes appear to be unusual in having such high rates of recent transfer, perhaps because we have been exposed to so many different new environments as we have spread across the planet. Intense effort is being devoted to understanding the effects that the multitude of species living in our personal microbiomes have on us—how they affect our uptake of nutrients and influence our susceptibility to disease.

Because of horizontal gene transfer, members of all microbiomes can share super-advantageous genes that otherwise might be trapped in one of the non-recombining genomes of the species in which they first appeared. It is therefore not surprising that microbial communities can evolve so quickly, as these transfers allow them to take advantage of new ecological opportunities.

How quickly can such changes take place? Garud et al. [88] followed changes in the genomes of forty different common species of bacteria in the gut microbiomes of individual humans over a period of months. Relatively straightforward genetic mutations and exchanges between the chromosomes of a species accounted for the majority of the changes, but a few horizontal transfers between different species could be detected even during that short time frame.

Horizontal gene transfer can prepare a population for changes even before the changes happen. Cagla Stevenson and her colleagues [89] mixed a strain of the bacterium *Pseudomonas aeruginosa* with a plasmid (a small piece of DNA that is found in bacteria but that is independent of the bacterium chromosome). The plasmid carried a mercury resistance gene. Strains of bacteria with and without the plasmid were grown for a while, and then the plasmid was removed and both types of strain were challenged with mercury.

In strains that had not been exposed to the plasmid, a few mercury resistance mutations arose. They spread to confer resistance through the usual pedestrian process of natural selection. But in strains that had been exposed, many lineages had already received mercury resistance genes as a result of horizontal gene transfers from the plasmids to their hosts' chromosomes. These resistant lineages immediately swept through their populations when the addition of mercury revealed their resistance.

Horizontal gene transfer can supply genes in large numbers to such plasmid- or bacteriophage-infected populations, some of which could survive even though those genes might have been useless at the time they were supplied.

Additional evidence for the swift evolutionary impact of transfers has emerged from further work of Mike McDonald and his collaborators [90, 91]. They used the pathogenic bacterium *Helicobacter pylori*, which (as I can attest from my own inadvertent interaction with it) causes stomach ulcers by infecting the lining of the stomach. *Helicobacter* readily takes up fragments of DNA in the environment, and can often incorporate the fragments into its own chromosome.

McDonald's group began with a *Helicobacter* strain that was sensitive to the antibiotic metronizadole. They mixed this strain with DNA taken from a different strain of the same bacterium that was resistant to the antibiotic. Whole-population sequencing showed that even in the absence of the antibiotic, they could detect 40,000 cases in which the bacteria picked up bits of this introduced DNA and incorporated them. The introduced bits often lowered the fitness of their bacterium hosts, but genetic recombination within the bacterium cells themselves tended to remove the most deleterious of these gene combinations.

Antibiotic-resistant strains of these augmented bacteria quickly took over the populations when the antibiotic was introduced. Control strains that had not had access to the external DNA adapted slowly, if at all.

The augmented strains actually showed faster growth rates in the absence of the antibiotic than the original antibiotic-resistant strain from which the DNA had been derived. This suggests that various other combinations of the introduced genes may also have recombined with the recipient strain to produce bacteria with increased general fitness.

These and other exciting instances of the important and ongoing role that is played by gene transfer mechanisms are emerging from the work of many scientists worldwide. A recent review of the role played by microbiomes in the adaptation of coral reefs and other complex ecosystems, by Christian Voolstra and Maran Ziegler of Germany's Justus Liebig University [92], concentrates primarily on how microbial communities can shift in numbers to accommodate themselves rapidly to changing environmental conditions. But the authors also point out that gene transfers among these microorganisms are likely to play an important role.

These authors emphasize the immediate changes in adaptability that microbiomes confer. In my discussions here, I have tended to emphasize longer-term evolutionary changes. Both types of evolutionary change, of course, are central to the adaptability of the world's evolutionary cauldrons.

In the next chapter, we will see some examples of how such horizontal transfers have made possible startling new directions in evolution, including the emergence of multicellular life. We will see how multicellular species like ourselves have benefited in many ways from the immense opportunities provided by such gene transfers.

The microbial world is a huge evolutionary resource, but it is not the only one that evolution can draw on. As we explore real-world examples of adaptation, we will begin to see how these microbial communities can also aid more genetically constrained species like ourselves that do not swap genes with such sheer abandon. Gene transfers have aided us, and undoubtedly continue to aid us, to survive in and adjust to our rapidly changing world.

8

Benefiting from the Bubbling Evolutionary Cauldrons

This discovery, indeed, is almost of that kind which I call Serendipity, a very expressive word, which, as I have nothing better to tell you, I shall endeavour to explain to you: you will understand it better by the derivation than by the definition. I once read a silly fairy tale, called "The Three Princes of Serendip;" as their Highnesses travelled, they were always making discoveries, by accidents and sagacity, of things which they were not in quest of…

Letter from Horace Walpole, Earl of Orford, to Sir Horace Mann, 1754

What are the real roles of evolution in an entangled ecological web such as Darwin's bank? Darwin's prediction that interspecies interactions are primarily responsible for such entangled ecosystems immediately opens up an array of possible answers. In this chapter, we will explore some of them, using examples from both the microbial and the multicellular worlds. These examples illustrate how evolution has dealt with some of the immense challenges of the past.

All living species are embedded in vast ecological and evolutionary entanglements. But we humans are an outlier among them because most of the ecological dangers facing us are the result of our own activities. Fortunately, we have evolved the ability to cooperate consciously with each other in order to overcome our self-generated problems—provided that we choose to do so.

Most of the species that are members of entangled ecosystems do not have this option. Their evolution is dependent on their opportunities for survival and reproduction, which as we have seen are mainly shaped by the other species with which they share the entanglement. These interactions can be cooperative or competitive. But cooperative interactions will only be selected

Figure 8.1 Edward Hicks' "Peaceable Kingdom."

for if the cooperating organisms, by acting together as a unit, become better utilizers of their environment or better at competing against the swarms of other species with which they share their entangled bank. Competition limits the extent of cooperation. Those delightful depictions of the Peaceable Kingdom by nineteenth-century American artist Edward Hicks (Figure 8.1) are, alas, unachievable in the real competitive world.

Both competitive and cooperative types of interaction play important roles in such an entangled bank, and both can take many evolutionary directions. In this chapter, I will show how these paths can contribute to the multidimensionality of ecosystems, sometimes in remarkably subtle and unexpected ways. Let me begin with some striking examples from multicellular organisms.

Hiding in Plain Sight

Sometimes, if an ecosystem is old and stable so that some of its entanglements have settled into predictability, its species have the time to "fine-tune" their responses. This is done through a long process of natural selection, in which even slight deviations from a particular body shape or behavior can be selected for if they enhance survival. Such processes of fine-tuning have produced some of the living world's most striking, and most delicately balanced, examples of evolutionary adaptation.

In 2006, I came across one such adaptation in Ranomafana National Park in east-central Madagascar. This wondrous rainforest is a showcase of unique and highly endangered ecosystems. Ranomafana provides a home for many species of lemur, tenrec, and chameleon, along with unique palms and orchids and other amazing organisms.

During a hike into the heart of the forest, I spotted a leaf-tailed gecko, crouching motionless in a bush that overhung the trail. Many geckos imitate features of their immediate environment in order to stay undetected as they wait for prey. The body and tail of this gecko were, with astonishing fidelity, imitating a large, curled, dead leaf.

My picture shows some details of the masquerade. The gecko's body and tail are clearly patterned in ways that imitate the midrib and lateral veins of a leaf. The tail curls back as if the sun had dried it. Black spots on its body, along with apparently nibbled places along the margins of its tail, seem to result from attacks by caterpillars. White spots resemble the growth of fungi.

The only feature that misses the mark is the gecko's overall color, which does not quite match the color of the real and freshly dead leaf that it is resting on. And that is probably why I was able to spot this shy creature in the first place.

Consider what must have happened in order for this remarkable disguise to evolve. Many genetic pathways had to be tweaked to produce the overall tail and body color, shape, vein patterns, and various types of spot. The apparently nibbled areas in the margins of the tail are indeed real holes, the edges of which have been shaped by natural selection to match nibbles that hypothetical caterpillars might have made before the leaf died. There are a moderate number of such features on the gecko's body and tail, not too many and not too few. And, if you look closely at the "nibbles" along the edges of the tail, you can see that they show tiny irregularities such as might have been left behind by a caterpillar's mouth parts.

Remember that no caterpillars have nibbled at the gecko's tail, and no fungi have invaded its tissues. These features have not been made by external forces—their pattern is the result of developmental programs that have, step-by-step, become built into the gecko's genes.

The amount of fine-tuning involved in this process is astounding. Infinitesimal differences in appearance, such as the difference between tails with semicircular nibbles that resemble those made in leaves by your typical caterpillar and other tails that have more squarish and therefore slightly less convincing "nibbles," must have yielded a slightly greater and a slightly smaller probability of fooling predators.

The geckos' behavior too has been similarly shaped. Some animals in the population will be better than others at balancing the complex tradeoffs between the safest places to sit motionless and the best places from which to watch for prey. These animals will have a minuscule advantage over those that are slightly less sensitive to the relevant array of environmental cues.

Long histories of continual testing by natural selection have resulted in the persistence in the geckos' gene pools of genes that influence these phenotypic features and behaviors.

Such an obsessive-compulsive evolutionary process is only possible in a stable ecosystem, in which many of the immediate interactions between a species and its various predators and prey undergo little change. But, while the geckos living in this stable Madagascar rainforest were evolving a beautifully adapted suite of characteristics to avoid large predators, other challenges to their survival were undergoing rapid shifts.

Defenses against Invaders

Some of the most quickly evolving challenges in an entangled bank ecosystem come from the interactions between host organisms and their parasites and pathogens. Disease organisms and parasites have been attacking their hosts almost since the beginning of life. We saw, in Ichihashi's experiments, how quickly "parasitic" RNAs can evolve in even the simplest of laboratory ecosystems.

Pathogens and parasites present a continually evolving threat to their victims. And host–pathogen interactions can sometimes leap to entirely new modes of warfare. Such interactions generate very different evolutionary challenges from those that were experienced by that gecko species as it was being tweaked by selection for a better disguise. Geckos had the leisure to fine-tune their pretend caterpillar nibbles over many millennia. Challenges by pathogens and parasites are far more immediate and change more rapidly. Host organisms that survive these shifting demands must be able to participate in an endless, rapid, evolutionary race between new defenses and new modes of attack.

How have the potential victims of pathogens in an entangled bank managed to keep abreast of such complex shifts in their attackers? Sometimes by out-flanking them.

In 1989, in a book called *The Wisdom of the Genes* [93], I began with the obvious. All lineages of organisms are sure to have had a long history of being confronted repeatedly by challenges posing threats to their survival. The threats may be from the physical environment, such as droughts, ice ages, or alterations in the chemistry of the oceans or the atmosphere. Other threats—over time, probably the vast majority of them—come from the many other species with which this particular lineage shares its ecosystem.

My thesis in that book was that the evolutionary lineages that are most likely to have survived such threats have done so by evolving systems of genes that can do more than simply respond to each threat *seriatim*. The most successful of these lineages, I suggested, would have evolved an unusually resourceful way of finding solutions to such problems, along with a high probability of mutating in ways that increase the chance that this resourceful system will become even more resourceful over time.

This is not a case of evolution somehow anticipating the future. Properties of genes, and of interacting systems of genes, evolve as a result of Darwinian natural selection, which sorts out random mutations that have arisen over time. Evolution cannot see ahead in time to deal with threats that have yet to happen. But there are many ways to evolve a defense, some more flexible and versatile than others.

Serendipity in the Evolutionary Driver's Seat

In the famous Monty Python sketch, nobody expects the Spanish Inquisition. In real life, nobody expects the arrival of an asteroid. But it is possible for organisms to evolve remarkably flexible defense mechanisms against less unpredictable but still dangerous threats. Such defense mechanisms may, because they have been selected for versatility in the face of continually varying challenges, have acquired the ability to go beyond formulaic responses to the challenge of the immediate moment—the challenge *du jour* if you like.

The evolution of such complex and subtle mechanisms is not an everyday occurrence. But if certain features of an especially versatile protective mechanism were to appear in a host species' gene pool, through a combination of mutations, genetic recombination, and selection, then these more versatile features would be the ones most likely to survive subsequent blizzards of challenges.

The slow construction of evolution's most elaborate and multifaceted adaptations depends on the accumulation of innumerable such rare happy accidents. It is these processes, not some supernatural ability to see the future, that I called "the wisdom of the genes." As Darwin had realized, natural selection is capable of producing things that can strike us dumb with wonder.

A superb example of the genes' wisdom is the evolution of the adaptive immune system. This is the cellular defense mechanism that we, along with all other animals with vertebrae and jaws, chiefly rely on as a shield against invading pathogens.

The COVID-19 epidemic has reminded us how much we depend on this highly flexible part of our immune system. Without it, we would be far more susceptible to invading bacteria, viruses, and other nasties. People who lack a functioning immune system, and people whose immune system has been suppressed for therapeutic reasons or because of genetic defects, are at immense risk from invading pathogens.

The adaptive immune system dates back to the time of the first appearance of jawed fish, which recent discoveries of fossils in China have pushed back to at least 485 million years ago. The system's sheer beauty and elegance make its evolution through a series of serendipitous accidents especially astounding.

Luckily, we now have many pieces of this puzzle, some of which date back to long before the adaptive immune system evolved. Some important DNA-rearranging genes that play a central role in the function of our immune system had actually arrived in our early ancestors' genomes through horizontal gene transfers hundreds of millions of years before the appearance of the first jawed fish. These genes resemble other DNA-rearranging genes that are often found in plasmids in present-day bacteria.

A long-vanished marine ecosystem must have brought together a virus or other transfer agent and an early animal, setting in motion what would become a long series of serendipitous events. We infer that this is what happened because sea urchins and other animals that are very distantly related to us, which do not have an adaptive immune system, retain more complete copies of the infectious agents.

Those strange alien genes still exhibit some of the DNA-rearranging abilities of the agents that originally brought them. The surviving bits of those ancient pieces of DNA are now known as RAG1 and RAG2, standing for Recombination Activating Genes 1 and 2.

In addition to RAG1 and RAG2, our current adaptive immune system is the product of numerous other multi-step evolutionary events. Together, these have resulted in a highly specialized collection of immune system genes and the cells that express them.

Among these cells are B lymphocytes. These cells can make a wide range of antibodies that are able to bind to invaders' proteins and chains of sugars on the surfaces of their cells.

At any given time, largely by chance, a few of the antibodies made by a host's adaptive immune system can bind strongly to the surface of new infectious agents, even though they may never have encountered these agents before. These coatings of bound antibodies act as signals to invader-eating phagocytic white blood cells and other defensive cells that the new intruder should be engulfed and destroyed. If the invasion continues, additional cellular mechanisms direct the most successful of these antibody-producing cells to multiply quickly.

If the host has not encountered the invader before, these processes take some time—an unavoidable delay that may put the host at risk. We humans can now use technology to cut to the chase.

We can make harmless variants of the pathogens, or harmless molecules that have been isolated from them, and use these products to induce specific antibodies in individuals who have yet to be infected by the real thing. We can train our immune systems (and those of our pets and domestic animals) to make antibodies that are ready to deal effectively with a pathogen, even before it actually attacks us. But when we do so we are only harnessing the resources of a defense system that has been half a billion years in the making.

There are two obvious possibilities for how such a system could have evolved. The first is that, under the growing onslaught of pathogens, the original antibody genes in the chromosomes of all our cells could have duplicated again and again to produce many copies. Such duplication events, which can arise from errors during chromosomes' replication and genetic recombination, are quite common. Over time, these new copies would each accumulate their own mutations, making each duplicated gene copy unique.

Each of these antibody genes would be able to make only one kind of antibody. But pathogens, like the Greek river god Proteus, can take many forms. As pathogen threats continued to proliferate and evolve, our ancestors would have needed thousands of such duplicated genes to make a sufficient diversity of antibodies. And, if more of these genes accumulated on our chromosomes, they would have presented an ever-increasing target for additional random mutations that could destroy them. Such inactivating mutations would reduce the ability of their carrier to respond to a threat.

Fascinatingly, exactly such a system, which exhibits exactly this vulnerability, has evolved in another part of our genomes. Many animals, including those that live in the ocean or in lakes, depend for their survival on the ability to detect molecules in the air or water that surrounds them. These molecules can indicate danger, the possibility of food, or the availability of potential mates.

In vertebrate animals, this ability is provided by a large number of duplicated, highly similar, olfactory receptor genes in their genomes. Each receptor protein, made in cells of the olfactory epithelium, binds to one or a few of the wide range of odorant molecules in the environment. The signals sent by the olfactory cells as they bind to members of the odorant mix allow animals to distinguish among them.

This system has jaw-dropping capabilities. For example, trained dogs can alert physicians to patients who have lung cancer, with a high success rate even for early stages of the disease [94]. The possibility that dogs can detect other types of cancer is being actively investigated. Even the most highly trained human chefs and oenophiles are olfactory klutzes by comparison. And there is a good genetic reason for our shortcoming.

We humans, and our close relatives the apes and monkeys, depend strongly on vision rather than smell and have only about 400 functional olfactory receptor genes in our genomes. We are also burdened with an approximately equal number of receptor *pseudogenes*, which have undergone mutations that inactivate the gene.

These pseudogenes, which are now useless, will eventually be lost by subsequent mutations that delete them from the genome. But such deletions, because they must occur through random mutations, are unlikely to happen immediately. This is why the corpses of these inactivated genes still litter our genomes.

Intriguingly, dogs are able to accomplish their amazing feats with a mere 800 functioning genes. They also have pseudogenes, but only about 300. This suggests that the dog genes that still work have survived through strong selection for functionality. Rats, the current champions in the olfactory receptor competition, have 1,200 active receptor genes. But even rats have 500 pseudogenes, meaning that their ancestors have lost some of their receptors through random mutations.

Clearly, gene-disabling mutations place a strong limit on the number of duplicated genes that a gene pool is able to maintain, even when these genes code for an important function.

The adaptive immune system has followed a different, highly successful path, a path that has allowed it to minimize the consequences of incapacitating mutations. The system allows each of us, during our lifetimes, to produce an enormous array of different antibody molecules—far more than we could manufacture if every antibody had to be specified by its own gene. This is because the relatively few antibody genes that we inherit from our parents have been passed down to us in the form of linked collections of little bits of gene.

Each of these bits of antibody genes have themselves been duplicated a number of times, and the duplicates are grouped into clusters that are carried by a few of our chromosomes. These little bits can be assembled in many different combinations to make complete antibody genes, and the RAG1 and RAG2 proteins play an essential role in this process.

We do not inherit complete antibody genes. Instead, we inherit these little clusters of gene segments. Each of the segments is slightly different from the other segments in the cluster because of accumulated mutations. The segments provide Lego-like building blocks for antibody genes (Figure 8.2).

Like the odorant receptor genes, these clusters of gene fragments are passed down through the cell generations. All the cells in our bodies have them. But most of our cells cannot activate the RAG1 and RAG2 machinery that constructs antibodies by piecing together bits of the clusters. This activating ability is confined to the B-cells in our bone marrow.

RAG1 and RAG2 gene products, the enzymes coded by those ancient genes that arrived in our animal ancestors in the remote past, now play a central role

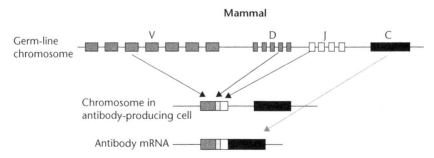

Figure 8.2 Different combinations of one each of V, D, and J gene fragments are joined together with a copy of the C region to make a complete antibody messenger RNA.

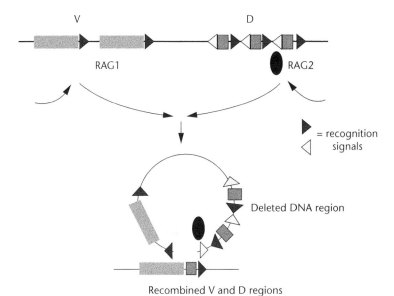

V

D

RAG1

RAG2

= recognition
signals

Deleted DNA region

Recombined V and D regions

Figure 8.3 How RAG1 and RAG2 delete intervening fragments of the antibody gene clusters to bring bits of an antibody gene together in the B-cells.

in our adaptive immune system. They have the ability to bring the pieces together in the right order to make complete functional genes. And these freshly constructed genes, unlike our other genes, are different in each of our B-cells (Figure 8.3).

Think of an Italian restaurant where (if you are sufficiently ravenous) you can choose one item from the *insalate*, one from the *antipasti*, one from the *primi*, one from the *secondi,* and so on to make a complete meal. If the restaurant's menu is sufficiently long and varied, it might easily happen that on a given evening, each of the customers assembles a meal that is unique.

Armed with a restaurant-like ability to make their own unique antibodies, each of our B-cells constructs its own antibody molecules with which to confront the hair-raising collection of pathogens that we will meet during our lives.

These fragments provide a surprisingly small target for mutations. For example, in humans, there are only three important clusters of the fragments carrying the code for the heavier of the two subunits of antibodies. The first cluster has sixty-five copies, the second has twenty-seven, and the third has only six. As with the olfactory receptor genes, there are also a number of inactivated fragments, but these do not play a role in functioning antibodies. Together, the functioning fragments are capable of generating 10,530 different possible combinations of the first, second, and third type of fragment. Light chains, with only two variable regions, can form a smaller number of possible combinations.

This does not seem like many possibilities. But the two heavy and the two light chains are formed at different times during the lifetime of the B-cell, and the heavy and light pairs can come together in different pairwise combinations to form complete four-chain (two heavy and two light) antibody molecules. This yields more than three million possibilities. And another mechanism adds even more variability to the antibody repertoire.

This mechanism is one of hypermutability, which also turns on only in B-cells and which acts on only the variable regions of the antibody genes. These genes affect DNA repair, and the result is a high rate of random mutations in these already-variable parts of the antibody. Together, these recombination and hypermutation mechanisms can produce billions of different antibodies. And the hypermutability mechanism, as we will see, has its own remarkable properties.

When we die, all our B-cells with all their antibodies die too. Any disabling mutational changes that might have happened in any of our B-cells during our lifetimes will disappear. It is only the much smaller sets of molecular Lego building blocks in our *germ-line* cells that we pass down to the next generation, along with the genes that are needed to put them together in different combinations. Over the long term, this mechanism has proved to be extraordinarily robust and effective.

This adaptive immune system, which seems at first sight to have been designed by Rube Goldberg, is far more flexible than the odorant receptor system. We all have slightly different collections of antibody gene fragments. As our pathogens come and go, and as they evolve during our lifetimes, those of us who are able to recombine our fragments to make the most effective set of antibodies in order to meet these shifting challenges will be more likely than other members of our population to pass on such an especially adept fragment set.

Why did our immune systems not take the more pedestrian gene-duplication approach that was followed by our olfactory receptor genes? Perhaps many lineages of animals did take that route, but were driven extinct because the world of pathogens presented such overwhelming challenges. It was those ever-shifting challenges that provided the opportunity for selection for such rarely occurring but highly effective new mechanisms that added versatility and slowly drove the evolution of the present-day adaptive immune system.

Clues to this history are being uncovered through the use of Darwin's principle of continuity (Chapter 1). Immune systems that are slightly different from the one that we jawed vertebrates possess have survived in other groups of vertebrates.

Sharks, for example, are on a different evolutionary path from bony creatures like us. They are, of course, famous for their jaws, but their skeletons are cartilaginous rather than bony. And they have a different arrangement of antibody gene fragments in their genomes from that of jawed bony vertebrates [95]. They do, however, use the same toolbox of alien genes that we employ in order to string those differently arranged antibody gene fragments together.

Slightly further afield, the even more ancient lineage of jawless fish (represented today by the hagfish and the lampreys) has evolved a very different kind of antibody [96]. These antibodies are much less complicated than ours. They are made up of variable numbers of subunits that are more similar to each other than the different fragments of our own antibody molecules.

These subunits are pasted together in the jawless fishes' antibody-producing cells by mechanisms that are not yet understood, but that do not involve RAG1 or RAG2. Infection causes an increase in the number of these antibody molecules, but apparently without the precise specificity of our adaptive system. This immune system seems not to be as versatile as ours, but perhaps it seems so because we do not yet understand it fully.

Compare these complex and branching evolutionary histories to the much less complicated evolution of the olfactory receptor gene family. One might be tempted to suppose that the ability to distinguish smells, even though it is sometimes a life-or-death matter, is not quite so critical for survival as the ability to resist diseases. This might explain why the olfactory receptor system, with its hundreds of duplicated genes, has not been replaced by a more flexible system in spite of the continual damage from mutations that disable some of those duplicates.

This hypothesis has yet to be tested, but the side-by-side existence of the olfactory receptors and the adaptive immune systems in a single organism shows that there is more than one way to survive the challenges of an entangled bank ecosystem. Especially rapidly changing and severe challenges may demand unusual sophistication in the systems that the host has evolved to defend against them.

Jawed vertebrates are unique in their sophisticated approach to defense against pathogens. But groups of animals that do not have jaws or backbones have evolved various elaborations of a more ancestral system, the innate immune system (one that we also possess). This innate system is unable to produce finely tailored defense molecules such as antibodies, but it does produce a wide variety of chemical and cellular weapons that can recognize and attack cells that are different from those of the host.

Perhaps it was just chance that a true adaptive system did not evolve in invertebrate organisms. They certainly have the potential to do so, because many of them carry different forms of those alien RAG1 and RAG2 genes that have played such an important role in the evolution of the jawed vertebrate immune system.

The alien genes are presumably doing something else in the invertebrates. Lina Carmona and her co-workers have shown that a RAG1 protein isolated from sea urchins can interact with RAG2 to carry out DNA rearrangements [97]. It may be that the ancestral mobile element was co-opted into carrying out DNA arrangements of some kind soon after it first invaded animals.

While there are some exceptions, invertebrate animals tend to have shorter lives and far higher rates of reproduction than vertebrate ones. Even the most complex of invertebrates, the octopuses and squid, live surprisingly brief lives—a maximum of five years for most species—while still managing to lay many eggs. Their short generation times and high fecundities may reduce the impact of diseases on their populations.

Ironically, our acquisition of highly sophisticated defenses has almost certainly pushed our own sets of pathogens to evolve ever-more-devious modes of attack. We and our pathogens are involved in what may be the most challenging and complicated set of host-pathogen arms races on the planet. But we have fought them to a draw, and perhaps to better than a draw.

In the future, were a highly effective adaptive immune system to evolve in invertebrates, it might help some of them to grow in complexity, longevity, and brain size, and to take over the planet from the vertebrates.

Fiddling with Evolution Is Sometimes Permitted

I suggested earlier that the adaptive immune system is a superb example of the wisdom of the genes, the ability to become better at overcoming challenges even though those exact challenges might not have been encountered before. Our genes themselves (and genes that we have borrowed) are often the source of these remarkable abilities. And our very evolution might not have been possible without the genetic resources provided to our distant ancestors by the microorganisms in the entangled banks in which they were embedded.

Genes are more than just passive stretches of DNA that undergo totally random mutational changes. They have been shaped during the course of their evolution in ways that make it more likely that, even when mutations to the gene happen at random, they will show a tendency toward changing the gene in particular directions.

For example, one of the most remarkable features of our adaptive immune system is the way in which we jawed vertebrates have managed to co-opt the process of mutation itself. We have evolved mechanisms that regulate mutation's level of activity in certain sets of genes of certain cells, while at the same time preserving mutation's random aspects. The result is the generation of narrowly focused new suites of mutations in parts of the antibody molecules.

This remarkable ability has arisen through the evolution of a process of localized somatic hypermutation that I mentioned earlier. The result of this process is a substantial increase in the rate of mutation of the variable regions of the immunoglobulin gene clusters, but only in the antibody-producing B-cells. These mutations are not passed on to the next generation.

Adding to the versatility of the Italian restaurant menu approach that generates different antibodies by mixing and matching pieces of antibody genes, hypermutation harnesses the process of random mutation to produce even more kinds of antibodies. Imagine that our Italian restaurant's waiters are encouraged to add a variety of toppings at random to each dish.

Note that our B-cells are not playing hanky-panky with evolution. Hypermutation is not a magical way of making specific changes to generate more effective antibodies. Careful studies have shown that hypermutation produces mutant antibodies that are sometimes more effective and sometimes less effective than the average, and sometimes indistinguishable from the originals. And sometimes the mutant antibodies do not work at all.

This is just what the more conventional sources of random mutation would produce, but at a higher rate. And an increase in the mutation rate, without loss of its property of randomness, is exactly what the system requires. Hypermutation simply increases the mutation rate, without restricting the boundless evolutionary possibilities that are opened up by the process of random mutation itself. If hypermutations of only a few types of mutation were permitted to happen, this would greatly restrict the variety of possible antibodies.

Without breaking the laws of evolution, we jawed vertebrates have evolved ways to enormously increase the available pool of genetic variation—but only in tiny regions of certain genes in a small subset of the cells in our bodies. Hypermutation provides this variation at exactly the time and place where natural selection can yield the most effective antibodies to deal with unexpected enemies. If a bacterium from another planet were to invade our world, even though its organic chemistry would likely be quite different from ours, we could probably make antibodies that could bind to it.

Evolutionary Breakthroughs

In these examples, we have explored some of the ways in which entangled-bank ecosystems can take on a kind of life of their own. Parasites push hosts to evolve. Hosts push back, forcing parasites to adapt in turn. A symphony of scents drifts through the air or diffuses through the water, signaling food, danger, or sex. Different species evolve new ways to prey on each other or hide from each other. These battles result in ecosystems that are continually adding more ecological niches, allowing new species to join the party.

As we saw in Chapter 3, ecologists are arguing over whether there is some upper limit to all this diversity. There probably is—at least in the short term.

For example, the effects of the hypermutations that are induced in a jawed vertebrate's B-cells all disappear when the animal dies. None of these changes ends up in the store of genetic variation in the species' gene pool.

Suppose that the genes that turn on the process in B-cells were to mutate so that they can also turn on in our germ-line cells. This germ-line hypermutation would be able to increase the diversity of the antibody gene fragments in the germ-line cells that are passed on to the next generation.

But this might be too much of a good thing. Many of these germ-line hypermutations would destroy the germ-line gene fragments, permanently removing them from the gene pool.

Such destruction must happen in the B-cells, but it does not matter because there are so many B-cells that the loss of a few of them is not a big deal. In our germ-lines, it seems that the regular gentle patter of ordinary new mutations can produce a sufficient diversity of antibody fragments to continue providing succeeding generations with a versatile protective library of these molecules to draw on. We jawed vertebrates can then fine-tune this diversity further, by employing somatic hypermutation to generate molecules that can deal with the latest challenges.

This ability may not be enough in the future. The AIDS HIV-1 virus exploits a weakness in our adaptive immune system. It enters and kills a class of white blood cells, the CD4 cells, that form an essential part of the system. This disables the entire system, and leaves the victim utterly at the mercy of a wide variety of pathogens.

Most AIDS-related deaths are the result of other infections or diseases— tuberculosis, pneumonia, viral hepatitis, and a wide variety of cancers that can spread easily in immunocompromised patients. Mortality from these multiple causes has reached forty million, almost half of the total number of known AIDS infections and a far higher number of deaths than those caused by COVID-19.

In most of us, our adaptive immune system is defenseless against this virus. The AIDS pathogen has been able to circumvent almost half a billion years of evolution. But some of us do carry genetic defense mechanisms.

In some human populations, a mutation has produced a tiny deletion in a gene called CCR5. The intact gene produces a protein on the surface of those virus-susceptible CD4 cells that is essential for the binding and entry of the HIV-1 virus. People who carry two copies of the deletion mutation, one from their mother and one from their father, cannot make the protein that permits the HIV-1 virus to enter the CD4 cells [98]. This mutation stops the virus from killing these white blood cells and damaging the host's immune system. Most, but not all, strains of the HIV-1 virus are rendered helpless by this mutation.

The mutation in CCR5 may also confer some resistance to the smallpox virus, and seems to have no obvious harmful effect on the white blood cells in which it is expressed. It is found in only a few of the world's populations. Its highest frequency is seen in northern Europe and central Asia, where it can reach a high of 15 percent of the alleles at this genetic locus [99]. These populations had certainly never encountered the AIDS virus until the current epidemic.

The most likely pathogen that originally selected for the mutant CCR5 is probably a virus, different from HIV-1, that was able to infect these populations but that subsequently disappeared or fell to low levels. Sequencing studies of the CCR5 mutation and two other markers that are closely linked to it suggest that this little deletion first appeared in the human gene pool approximately 700 years ago. Although its origin was recent, it has already become so common in Europe that it has very likely been strongly selected for.

Other mutations that prevent the infection by the HIV-1 virus of these essential white blood cells have been discovered more recently. One of them, TNPO3, has a huge cost that dwarfs the benefit conferred by its HIV-1 resistance. It is responsible for a form of the hereditary wasting disease muscular dystrophy [100], which explains why the TNPO3 mutation is extremely rare.

Our gene pool also carries a few variants of other genes that are able to confer resistance to this retrovirus. As one would expect, these resistance genes, the result of random mutations in the past, are a mixed bag. Still, as HIV-1 continues to spread, it is likely that the most effective and least harmful of the resistance genes will continue to spread as well. Thus, even if AIDS had appeared before the advent of modern medicine, we might eventually have evolved new modes of resistance, including new capabilities of our adaptive immune system and new mechanisms for CD4 cells to resist invasion.

Luckily, drug treatments are now available to mitigate the effects of HIV-1. Without those treatments, the capacity of this virus for damage and death would have remained largely unimpeded until increased frequencies of resistance mutations in the human population eventually brought the epidemic to an end—or at least to low levels.

I would suggest that if complex defense mechanisms are already in place, this increases the likelihood of—to use a loaded term—such an evolutionary breakthrough. And, as we have just seen, the possibility of such evolutionary breakthroughs can change the entire dynamic of interactions between species.

Simple and Complex Red Queen Races, and the *Status Quo Ante*

The simplest host-parasite competitions are evolutionary arms races, which are just like human arms races. One side evolves a new weapon and starts to win. Then the other side takes the lead by evolving a counter-weapon or an effective defense, and the cycle repeats.

In such simple arms races, new alleles of genes replace other alleles in turn in each of the competing populations. The result is a rather monotonous succession of replacements.

In 1973, paleontologist Leigh Van Valen suggested a wonderful and vivid analogy to explain this basic process of an evolutionary arms race. He introduced the analogy as a way of encapsulating what he felt must underlie not just

host-parasite interactions, but much of the evolution of competitive interactions between species and their biological and physical environments.

The analogy draws on Lewis Carroll's story of a race between Alice and the
Red Queen, which takes place in the magical dreamscape of Looking-Glass
Land. In Carroll's race, the Queen pulls Alice along and they run furiously
until they are exhausted. Alice is surprised to discover that they have not gone
any distance at all, though they would certainly have done so in her world. The
Queen contemptuously dismisses Alice's world as too slow, and explains that
in Looking-Glass Land, one must run as quickly as one can just to stay in the
same place.

Van Valen was thinking in very general terms when he proposed this analogy.
He suggested that much evolutionary change, affecting all aspects of the
lives of organisms, is driven by pressures of the Red Queen type, in which
groups of organisms must evolve as quickly as they can simply to keep up with
environment changes—of all kinds. He did not specifically emphasize the
circumstance with which the Red Queen has become associated, the necessity
of keeping up with the other species with which a species interacts because
those species are also evolving briskly.

Van Valen's analogy certainly describes what is happening in evolutionary
arms races, but I do not think that it describes adequately the kinds of interaction that we have been exploring in this chapter and that can produce unex-
pectedly versatile defensive and offensive mechanisms.

Initially, Red Queen races simply described desperate attempts to keep
abreast of the competition. As with Alice and the Red Queen, at the end of a
race, nothing has really changed and the participants are still confronted by the
status quo ante.

These days, and I think rather confusingly, the Red Queen concept is being
arbitrarily used to lump together simple arms races and the evolutionary races
that we have explored in this chapter and that have gone far beyond such simplicity. In those more complex races, as we have seen, strong selective pressures
can sometimes result in the emergence of utterly unexpected capabilities.

The subtleties of such interactions can be endless. They have produced such
dazzling mechanisms as hypermutability, which as we saw loads the evolutionary dice.

I think this broad use of the term Red Queen, without further qualification,
is misleading. Better perhaps to divide these races into simple and complex Red
Queen races.

Evidence is now emerging, from experiments involving simple laboratory ecosystems, that the first competitive interactions to evolve are classic arms races—simple Red Queen races. But these are quickly supplanted by more complex interactions.

A recent experiment by Alex Hall and co-workers [101] started with a population of *Pseudomonas* bacteria in which all the cells were initially genetically identical. They infected this population with one of the bacteriophages that prey on these bacteria. The phages, too, began as an initially genetically uniform population.

The two populations quickly began to evolve enhanced defenses and enhanced infectivity in turn, following the steps of a simple arms race between the bacteria and the viruses. Then, within a few hundred bacterial cell generations, the average effectiveness of successive defensive and offensive steps began to decrease. Instead, later changes tended to increase both the *variation* in defensive capabilities among the bacterium hosts and the *variation* in the ways in which the phage parasites could infect their hosts.

Such progressions from simple to complex Red Queen races are likely to be characteristic of all ecosystems. Even apparently simple planktonic ecosystems, as we saw in Chapter 5, have a complex mix of hosts and pathogens that contribute to their stability and that may enhance the role they play in our planet's fragile balance between chemical and biological processes.

The limiting factor here is time. A fine wine may take decades to achieve its full flavor and complexity. An awesomely versatile result of complex Red Queen races, such as our adaptive immune system, may have passed through hundreds or thousands of intermediate stages.

When ecosystems are damaged, they may lose more than just species diversity. As the number of surviving species decreases, ecosystems can lose their painfully acquired ability to draw on that diversity and adapt it to new conditions. But, as we saw throughout the immensely long history of tropical reefs, given a sufficient period of settled conditions, there is no doubt that a comparably sophisticated diversity will eventually re-emerge.

9

Venturing beyond the Red Queen

In tracing the origin of the lightbulb, historians Robert Friedel and Paul Israel compiled a list of twenty-two people who invented incandescent lamps before Edison even filed his first patent for a lightbulb.

David Burkus, *The Myths of Creativity*, 2013

As we have seen, competition between species in entangled banks has led to some truly amazing adaptations. And so has an equally important evolutionary trend, the emergence of *mutualism*, which is mutual cooperation between species. Cooperation need not be limited by the zero-sum aspect of many competitions, in which there are never-ending battles over limited resources.

On our way to exploring such new directions, we will encounter challenging evolutionary problems. Let me begin with a brief introduction to and exploration of one fascinating evolutionary possibility in particular. The possibility is this:

Together, the organisms of Darwin's entangled banks interact to form a complex multifaceted environment. Can such mixtures of species, without breaking any evolutionary laws, acquire new collective properties of some kind that increase the ecosystem's overall stability and the likelihood that it can survive environmental shocks? If so, what shapes might those collective properties take?

I will have to tread very carefully here. I am suggesting that interacting species, which mostly work at cross-purposes with each other, can produce such an entangled bank.

Darwin, and most evolutionary biologists since his time, based their arguments for evolution on the assumption that natural selection acting on individuals—selection with an immediate effect on whether an individual's offspring survive—is the primary driver of impactful evolutionary change.

But the idea that natural selection somehow acts on groups, such as the inhabitants of a tangled bank, to increase the fitness of the entire group is an old one. Its siren call has repeatedly lured scientists into a Sargasso Sea filled with the floating wrecks of previous "group selection" arguments.

One recent argument for group selection involves the role of altruistic behavior. Such behavior can evolve through kin selection, which may harm an individual but may benefit the individual's relatives. Or it can evolve through reciprocal altruism—behaviors that may benefit unrelated individuals—but only if there is a good chance that the recipients of the benefit will reciprocate the favor. It has been suggested that a group that has many such altruists might be a better competitor than one with fewer, because the altruism would lead to a sharing of resources and increase the fitness of the entire group [102].

Most evolutionary biologists have concluded that altruistic behaviors may (or may not) make *individuals* exhibiting them better competitors, but that these properties do not somehow make the *group as a whole* a better competitor.

After all this *sturm und drang* over group selection, individual selection continues to reign supreme as the premier driver of evolutionary change. But selection on individuals can have a wide range of impacts. Organisms of the same and of different species really do interact with each other, imposing selective pressures on each other. How do these interactions, which are primarily competitive, lead to ecosystem stability?

For example, how many parasites and parasites of the parasites does a host species carry, and how are these parasites and parasites of parasites affected by their hosts' numbers and their hosts' host's numbers? How do different forms of genes in each of the gene pools influence their species' behaviors? How do the different behaviors themselves evolve as the mix of species changes and the selective pressures that are shaping each species' gene pool shift? And on and on.

The wild card here, which constantly shifts the grounds on which these interactions take place, is natural selection itself. If plentiful physical resources, such as sunlight and nutrient-rich water, are available to an emerging bit of land in the tropical ocean, then, as we saw in Chapter 3, some kind of tropical reef ecosystem is likely to become established. But the structure and range of interactions that were exhibited by such a reef long before the Cambrian, when all life was single-celled and when there was almost no free oxygen in the environment, were far more limited than those that are available in a coral reef today.

Nonetheless, Precambrian reefs were stable, balanced ecosystems. A few stromatolite communities with similarities to those reefs even survive today in unusual environments such as Western Australia's Shark Bay.

An important part of the story of how such stable ecosystem communities have repeatedly emerged can be found in the evolution of mutualistic interactions.

The cooperation between *Cecropia* trees and *Azteca* ants that we met in the Introduction is typical of one type of mutualistic interaction. Entire suites of morphological, physiological, and behavioral characteristics have been selected for in these two species. In the process, the two have become a kind of super-species, able to survive better together than either could by itself.

In such a robust system, even a bit of gentle cheating on the side is tolerated. The ants cultivate sap-sucking insects to gain extra nourishment from their host trees, and this little peccadillo does not seem to damage the overall ant–plant relationship.

Let me emphasize once more that this is not a process of group selection. All of these mutualistic interactions emerged as a result of separate selection-driven changes in the gene pools of the ants and those of the trees. The evolving ants acted as *external* selective agents on the evolving trees, and vice versa.

Some of these mutualisms, which began between two or more species, have now mixed their genes together to the point where the component species' gene pools have blended into one.

Such mergings have taken place in entangled ecosystems throughout evolutionary history, and they include some of the most fundamental of life's processes such as the mutualism between nucleated cells and their mitochondria and plant cells and their chloroplasts. The mergings would have been far less likely to happen in a simple ecosystem that had few between-species interactions. Two species have become one, fusing the fates of both. But other species that share the ecosystem will still continue to take independent evolutionary paths.

The most remarkable of these "blending" mutualisms were so complex, and seem to have evolved through so many unlikely steps, that their crucial steps may have taken place only once. But the moment they did evolve, they began to transform the planet. Such abilities could not have emerged except through opportunities that arose in entangled ecosystems and that aided the evolution of the participating species.

Consider the evolution of photosynthesis. The first living cells may have appeared on the Earth more than three and a half billion years ago, less than a billion years after the formation of the Earth itself. In order for these *protocells* to get to a point of sustainability, they had to survive one near-disaster after another.

For example, no matter how rich a "primitive soup" of organic chemicals might have been present on the early Earth when life first arose, it would have been quickly exhausted as soon as early protocells began to dine on it. Luckily, some of those protocells evolved ways of using some of those abundant and energy-rich compounds and elements to construct additional molecules that they could use for metabolism and reproduction. And they evolved mechanisms to store and translate at least some genetic information that could help their offspring to manufacture energy-rich molecules for growth and metabolism, and to pass that information on to their own offspring.

The mechanisms for generating energy-rich molecules started with fatty acid-rich molecules, which may have been present in the primitive soup and which were available to form boundaries that separated these protocells from the outside world. Those crude boundaries would quickly evolve into cell membranes. Today, as much as half the mass of our cells consists of such membranes, which divide the cells into many compartments.

Available sources of energy at the time included molecular hydrogen, hydrogen sulfide, and ferrous iron. They provided the precious energy that was needed for the populations of protocells to survive long enough to be able to evolve additional, desperately needed features. These membrane-bounded early structuresevolved into more complex cells, able to extract energy from these molecules.

By a series of what were probably hair's-breadth escapes, some of these protocell lineages evolved pathways leading to the synthesis of more complex energy-rich powerhouse molecules such as ATP. These *electron transport* pathways, which are now found in all organisms, were powered by strong charge differences that the protocells were able to establish across the membranes that surrounded them. The charge differences could be used, by early protein or protein-like molecules that became embedded in the membranes, in order to carry out essential syntheses [103].

So ancient are these systems that their origins stretch back before the time of the ancestor that gave rise to all of today's living organisms. We know this because every creature that is living today possesses an already-elaborate version of electron transport. All of them also possess many other basic cellular mechanisms and processes, including RNA and DNA replication machinery, along with the mechanisms by which the genetic code is translated into proteins. The genes for all these abilities are currently shared across the entire tree of life.

This means that our *last universal common ancestor* (LUCA), the cell that gave rise to all of today's living organisms, must have been able to do all these things. It also means that these elaborate structures must have evolved before the LUCA.

Astronomers are, as yet, unable to see back to a time before the Big Bang that began the universe. They can find no trace of what, if anything, led up to the Big Bang. But evolutionary biologists are luckier. They can look at the structures of present-day versions of ancient proteins. They can then use these structures to extrapolate backward to what the ancestors of those proteins might have been like at the time of the LUCA. They can even make educated guesses about how those early versions of the proteins might have evolved during the time before LUCA.

How much time was available for all these amazing advances to evolve? We are not sure, because estimates of the time back to the LUCA vary. If, as seems likely, the LUCA lived perhaps 200 million or 300 million years after the first appearance of life, all the properties that the LUCA possessed must have evolved within that time frame. This would have required an amazing amount of evolution!

In 1998, pioneering evolutionary biologist Carl Woese, the discoverer of the bacteria-like Archaea—an entirely unexpected and gigantic new branch on the evolutionary tree of life—examined the question of how evolution could have produced life itself. He suggested that early organisms may have evolved across a broad front, and became capable of exchanging genes through various modes of horizontal gene transfer that were mediated by transposable elements and early viruses [104].

The most successful of these gene combinations happened to be brought together in lineages leading to LUCA. Through a series of such happy accidents, they became more and more effective competitors and could move into all the ecosystems of that early world. Eventually all the earlier lineages, some of which had contributed parts of the LUCA's suite of abilities, were driven extinct.

Woese suggested a scenario that he called "genetic annealing," in which different components of the LUCA could have evolved separately in various early branches of the tree of life. The components in the different branches would have been continually tested by natural selection as they were brought together in new contexts. Genes for the abilities that survived this demanding process were finally combined in the genome of the LUCA ancestor by a series of horizontal gene transfers and mutualistic interactions.

Such a process is rather like having a legion of severe editors with different requirements criticize multiple versions of a manuscript simultaneously, at the same time as they are exchanging randomly chosen pages from these different versions with each other.

Woese's hypothesis was largely dismissed at the time as highly unlikely. The dismissers have, I think, not fully appreciated the ability of natural selection to sort out multiple variants, including those that had their origin in different lineages. Genetic exchanges from many diverse sources must also have been greatly facilitated because they happened in the midst of early versions of Darwin's entangled ecosystems. In those venues, the lineages were easily able to exchange genes and to evolve both mutualistic and competitive relationships. Perhaps those early evolutionary cauldrons bubbled even more briskly than the cauldrons of today.

There is strong evidence that another set of genetic changes with equally profound consequences happened billions of years after the appearance of the LUCA. These more recent events clearly did take place according to Woese's scenario.

Eukaryotic ("true nucleus") cells possess a complex nucleus bounded by a double layer of membranes, providing a shelter for chromosomes that is unlike any in the bacterial world. The origin of these cells, found throughout the great branch of the tree of life that includes complex multicellular organisms such as ourselves, is still hotly debated. What is not debated is that those first eukaryotic cells must have arisen in a series of steps involving mutualisms between bacterial, archaeal, and perhaps virus contributors, just as Woese had postulated. These mutualisms arose from multiple gene transfers as well as fusions between different types of cells and engulfments of one type of cell by another. In the process, many species with different properties contributed to the eukaryotic lineage that has given rise in turn to to so many complex species.

Hints of these processes may still be found in some archaeal cells that have been discovered living on the ocean bottom near a deep-sea volcanic vent off the Icelandic coast [105]. The vent is called Loki's Castle, and the species names of this group, the Lokiarchaeota, are also drawn from the Norse pantheon. Their names evoke an ancient world in which magical events could happen.

Some of these Lokiarchaeota, incredibly slowly growing, have now been raised in the laboratory in experiments that required superhuman patience [106]. These cells divide every two or three weeks, compared to a mere twenty minutes for the bacterium *E. coli*. In the time that it takes to raise two Lokiarchaeal cells from one parent, a single *E. coli* cell that is given unlimited space and food could pack ten billion universes as big as our own with its progeny.

The investigators' incredible patience over years of culturing paid off with enough cells to see under the electron microscope and to yield sufficient DNA for sequencing. These Lokiarchaeal genomes carry some genes, and have cellular structures, resembling genes and structures that had previously been seen only in eukaryotes.

Details of those early Woesian annealing events are only emerging with exquisite slowness, requiring infinite patience on the part of the investigators. But we can follow the bubblings of more recent evolutionary cauldrons in more detail.

Even after the emergence of early cells that could make copies of themselves, immense and widely available sources of chemical energy, needing only light and carbon dioxide, remained beyond their reach. Any lucky organisms that could utilize these resources would be able to spread everywhere on the planet that light could reach.

Light was utilized first. Very early, through multiple steps that are mostly lost to history, some lineages of bacteria evolved complex collections of membrane-bound proteins, pigments, and chlorophyll molecules that together made up a *photosynthetic reaction center*. That center could channel light energy and use it to strip an electron from a donor molecule. Donor molecules were energy-rich molecules such as hydrogen sulfide, ferrous iron, or even molecular hydrogen, all of which were plentiful in parts of the anaerobic, volcano-studded early Earth.

The electron, as it traveled through various stages of electron transport, could power a series of other reactions giving rise to other molecules such as ATP that were essential for the cell. In sulfur-generating bacteria the electron "reduced" the hydrogen sulfide to elemental sulfur in the process (Figure 9.1).

An amazing glimpse of how this process might have unfolded through multispecies contributions was reported in 2017 by Phuc Ha and co-workers [107]. They found an anaerobic photosynthetic bacterium, *Prosthecochloris aestuarii*, that could photosynthesize by itself and that could also borrow high-energy

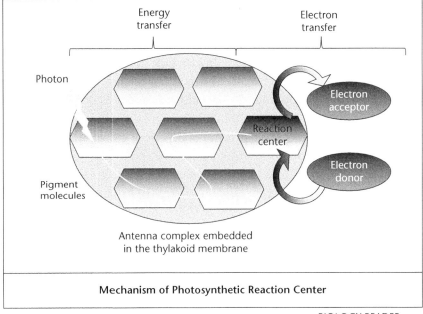

Mechanism of Photosynthetic Reaction Center

BIOLOGY READER

Figure 9.1 In photosynthesis, light photons reach an "antenna complex" and excite a series of pigment molecules that pass their energy to the chlorophyll-containing reaction center. This center strips electrons from an electron donor, such as hydrogen sulfide, and passes them to an electron transport chain. (Redrawn from a diagram used by *Biology Reader*, with permission.)

electrons from a non-photosynthetic "partner" bacterium, *Geobacter sulfurreducens*. The photosynthetic ferment among anaerobic bacteria is continuing, it appears.

Those early photosynthetic bacteria quickly dominated the ancient world. It was the activities of hydrogen sulfide-dependent photosynthesizers that generated the huge beds of elemental sulfur lying beneath the coast of the Gulf of Mexico.

These various bacteria were *anoxygenic photosynthesizers*, able to thrive because they could use light energy. But they were still confined to places where molecules such as hydrogen sulfide or ferrous iron happened to be plentiful. Free oxygen was not a product of those reactions. That transformational change had to await the evolution of the first *oxygenic photosynthesizers*.

What happened next set the stage for the evolution of a series of subsequent mutualistic interactions that must have followed a path similar to the one Woese suggested for the evolution of the LUCA.

As early photosynthetic bacterial lineages branched off from each other, they evolved different variants of their photosynthetic reaction centers. Each of these variants was well-adapted to a particular wavelength of light, and was able to use a particular energy-rich molecule as an electron donor. This diversity of abilities allowed photosynthetic bacteria to thrive in many of the environments of early Earth.

Entangled ecosystems emerged that included mixtures of such anoxygenic photosynthesizers. These mixtures, probably mostly in stromatolite communities, could maintain both the photosynthesizers and the other bacteria in the communities through a complex balance of competitive and mutualistic interactions.

One of these anoxygenic photosystems is Photosystem I, now found in the green sulfur bacteria. This photosystem can generate both the moderately energy-rich compound NADPH and the more versatile ATP. Another variant, Photosystem II, is found today in purple sulfur bacteria. Its only energy-rich product is ATP.

The quotation at the beginning of this chapter shows that Thomas Edison's success in harnessing the power of electricity to illuminate the modern world was preceded by many failed attempts by other workers. Edison's team beat out the others by harnessing a greatly improved incandescent light bulb to a greatly improved electric generator. But his team only succeeded because they could learn from those earlier attempts, and could utilize many other breakthroughs in our understanding of how to exploit and distribute electric power.

An Edison-like process of synthesis, competition, and teamwork took place among these early light-utilizing organisms as well. Somehow, the entire sets of genes for both of the two distinct early photosystems I and II were transferred into an entirely different bacterial lineage. This lineage had not itself been photosynthetic before these transfers. Once there, these sets of genes evolved the ability to work as a team to use light to split water instead of splitting hydrogen sulfide.

Splitting water and extracting an electron from it takes a great deal of energy: 260 kilojoules per mole of water. These new bacteria could perform this chemically challenging task with incredible panache. The blending of the two photosystems, accompanied by many further modifications, enabled these new photosynthesizers to perform feats of biosynthetic derring-do that no single lineage of organisms had ever accomplished before.

In the newly cooperating photosynthetic reaction centers of these bacteria, two energy-rich electrons were torn from water by light energy.

Both were passed up the steps of Photosystem II, where they used electron transport to power ATP production. Then the electrons were seized by Photosystem I and jacked up again to even higher energy levels, enabling them to power further synthetic work. Not only could these bacteria photosynthesize using water, but they could carry out their biosynthetic reactions better than ever.

These new oxygenic photosynthesizers, which would become the ancestors of today's cyanobacteria, seem to have immediately left their mark on the planet. Geochemical analysis has provided strong evidence for the appearance of oxygen in the Earth's atmosphere about two and a half billion years ago, long before the appearance of multicellular life [108]. Some of this oxygen could have been generated through non-biological processes, but it is striking that the consensus of times for when cyanobacteria-like lineages first appeared, estimated from rates of DNA divergence, turns out to be just prior to this first great burst of free oxygen [109].

The bacteria that were able to accomplish this transformation had a substantial impact on their almost oxygen-free world. They could live, not just in fetid swamps or on the slopes of sulfur-rich volcanoes, but anywhere there was air, water, and sunlight.

These early ancestors of the cyanobacteria continued to be resourceful. Some of them invaded single-celled eukaryotic organisms—which, as we saw, had themselves arisen from an earlier Woesian annealing that had combined bacteria, viruses, and Lokiarchaeota.

In their new home, the cyanobacteria were now trapped in their hosts' cells. Even though they had lost their identity as species, they were conferring immense benefits on those hosts. During their adaptation to their new intracellular life, they underwent great changes, losing many of their genes to the chromosomes of their hosts. Their descendants would become the chloroplasts, while their hosts would give rise to all of today's green plants.

But the low and possibly fluctuating levels of the first oxygen increases suggests that other factors were in play. Much of this free oxygen was quickly taken up by oxidizable deposits of iron. And, even as these early oxygenic photosynthesizers invaded new ecosystems, there still must have been opportunities for older lineages of anaerobic microbes to compete with them.

That early burst of oxygen was actually rather small, reaching only 1 or 2 percent of the atmosphere compared with today's 20 percent [110]. The next great increase occurred about 600 million years ago, fifty million years before the appearance of a large variety of multicellular organisms at the start of the

Cambrian. Oxygen levels climbed to about 5 percent. They reached their all-time peak during the forest-dominated Carboniferous. This peak began about 450 million years ago and reached an astonishing 30 percent or more of the atmosphere. This great increase may have enabled the dominance of immense forests that would give rise to the world's coal beds, and of many species of land-dwelling reptiles that would eventually become the dinosaurs, their descendants the birds, and the mammals.

The evolution of the first oxygenic photosynthesizers seems to have happened only once (though it probably came within a whisker of happening many times). The result was the world-changing system that produces the oxygen we all breathe and most of the food that we eat.

This trajectory is very different from the far more pedestrian process leading to the beautifully adapted leaf-tailed gecko that we met at the beginning of Chapter 8. In that case, mutations resulting in small changes in the geckos' appearance were repeatedly tested against selective pressures that were applied by predators able to use vision to locate their prey. The predators undoubtedly grew cleverer over time, but the geckos were able to keep up with them during that simple and narrowly focused arms race.

The evolution of oxygenic photosynthesis, in contrast, involved ventures into new, previously inaccessible, universes of adaptation. There is, I suspect, no way by which all these changes could have occurred in any reasonable time frame without the kind of evolutionary free-for-all proposed by Carl Woese with its many types of genetic annealing.

How wonderful it would be to be able to go back in time and visit (to adapt a line from the musical *Hamilton*) the stromatolite where it happened!

"Genetic Annealing," the Wisdom of the Genes, and Entangled Banks

In this chapter, we have dealt with a bewildering variety of evolutionary changes, some of which have altered the course of the entire living world. Are all such amazing evolutionary events confined to the distant past? Perhaps not.

We have come a long way from our starting point. We began with an exploration of the information that Darwin had used to form his original vision of an evolving entangled bank. These banks are turning out to be far more complex than he could have imagined. They harbor cauldrons of evolution at every level of biological complexity, and they can encompass every imaginable pattern of ecological and evolutionary interactions among their component organisms.

Science is now moving to the point where it is feasible to re-examine, and perhaps to recreate in the laboratory, today's entangled ecosystems. We cannot yet keep track of these ecosystems' entire web of evolutionary interactions or their full range of positive and negative selective pressures. But we can begin to glimpse the kinds of questions to ask.

For example, how resourceful is a given ecosystem at generating new competitive and cooperative biochemical mechanisms, including possibly game-changing mechanisms that exhibit unusually broad capabilities (the "wisdom of the genes")? How much does this ability depend on the number and types of species that are present—including different "species" of virus? How much do such advances depend on mechanisms of horizontal gene transfer among species—the Woesian genetic annealing free-for-all?

Time presses on us as never before in the history of life. Alas, the ecological damage that we are wreaking on the ecosystems that support us is taking place over spans of decades rather than millions of years. Truly new genetic mechanisms are unlikely to emerge during such a period unless they are on the cusp of doing so. Our species' growing ability to squeeze ecological destruction into a short span of time threatens to render moot our realization of the enormous consequences of failure. The collision between denial of indisputable facts and increasing comprehension of the facts' consequences adds a growing sense of urgency to our search for the secrets of ecosystem resilience.

All is not lost. I propose that many versatile genetic mechanisms may already be nascent in the swarms of gene pools in present-day entangled banks, though few may rise to such empyrean heights as the emergence of aerobic photosynthesis. It may take only a few serendipitous genetic accidents for some of these adaptations to be revealed, in at least their early stages.

It may soon become possible to observe, on a small scale, relatively dramatic evolutionary events taking place in the short space of time of a laboratory experiment or an attempted ecological restoration. What kinds of ecosystems would we need to start with? What kinds of challenges could we present that would allow the ecosystems to reveal their true evolutionary potential?

Using such approaches, will it be possible for us to measure and eventually predict the ability of such entangled ecosystems to respond to more severe environmental challenges? And what features of the entangled ecosystems might need to be modified or enhanced, in order to increase the chance that they will be able to respond?

Experimenters keep laboratory ecosystems simple, because otherwise the number of variables quickly gets out of hand. But the simplicity of these laboratory ecosystems makes the emergence of the kinds of far-reaching adaptations that we have explored here exceedingly unlikely. We cannot explore the true resilience and potential of Woesian free-for-all genetic annealings unless we throw all sorts of species, at every level of complexity, into the mix as we try to recreate the conditions that are needed.

The challenges are formidable. Consider Malaysian rainforests and the utterly depauperate ecosystem of the oil palm plantations that are replacing them. Elsewhere [6], I tell of flying over these plantations in peninsular Malaysia, and seeing them spread out below me like an immense pair of green corduroy trousers. Walking through such plantations is a depressing and dangerous experience—the understory seems to consist entirely of ferns, rats, and snakes.

This depauperate physical environment seems not yet to be reflected in the microbial environment. Soil microbiomes in the plantations are at least as diverse as in intact or recently logged forests [111].

It may be that tropical microbiomes are more tenacious than temperate ones. In China's Jiangxi Province, with a warm but more seasonal climate than in the tropics, microbiome diversity dropped precipitously after only three or four seasons of peanut crops grown in monoculture on land that had been fallow for more than a decade [112]. The decline was by a factor of three among common bacterial genera, but by a factor of more than twenty in rare genera.

What is happening in such stressed microbiomes? What are the chances that horizontal gene transfer can happen in stressed and unstressed microbiomes? How quickly can stressed and unstressed microbiomes respond to challenges of various kinds? These are all questions for future generations of ecologists.

It is possible that microbiomes are more resilient than we expect. In the next chapter, we will meet some of the highly stressed ecosystems that are ripe for such studies. The final chapter will explore some of the experimental possibilities that are presented by the world's evolutionary cauldrons.

10

Tipping Points

The Genghis Khan Effect

Entangled ecosystems nurture many between-species interactions that help to keep them stable and healthily diverse. And, as we just saw, these interactions help to power evolutionary changes in the thousands of species that are involved.

Long periods of time, along with complex exchanges of genes and shufflings among them, must have played a role in the evolution of many of the most important mechanisms that aid adaptation. If "wise" adaptations emerge during this process, their abilities can often be applied to new situations. When, for example, a species of mammal invades a new ecosystem, its adaptive immune system is already well prepared to be able to defend against new pathogens and parasites that it might encounter.

What happens to organisms, even those that possess such painfully evolved capabilities, when ecosystems spin entirely out of control? Can the organisms' capabilities cushion the blow? We must always remember that evolution is unable to prepare living things to survive the utterly unexpected. But it is also likely that the species interacting in today's ecosystems have become more resilient at dealing with at least some of the challenges that have unexpected aspects, compared with the species and ecosystems of the past. And if so, how can we best tap into and draw on this ability?

Our planet has no shortage of severely human-impacted ecosystems, In 2008, I found myself in such an ecosystem in western Mongolia. This remote country is bounded by China to the south and Russia to the north, and extends out to Kazakhstan to the west. Its westernmost part is dominated by the rugged Altai Mountains, with their glacier-fed lakes and dry high-altitude valleys.

This area has been occupied by human hunters and pastoralists since the time of the first great migration of modern humans out of Africa (Chapter 2). Over the last 10,000 years, the peoples of this region have left a vivid record of what life was like for them. The record takes the form of a kind of graphic novel. Clusters of remarkable petroglyphs, scattered throughout the region across many remote canyons and rocky hillsides, represent scenes that span much of this time period.

The most recent of these petroglyphs, some of which I visited in the Aral Tolgoi area, show mounted warriors guarding huge carts pulled by teams of horses. The petroglyphs also show lifelike depictions of local animals, including gazelles, Saiga antelope, musk deer, and wild boars (Figure 10.1).

Figure 10.1 Petroglyphs, from a time shortly before Genghis Khan, in western Mongolia.

These vivid glimpses of life in the area have not been precisely dated, but details of the horses' harness and of the warriors' weapons and armor provide clues. The ones that I saw were recent, appearing to date to the late medieval Turkic period around AD 1000. During that time, when Europe was still slowly recovering from the breakup of the Roman Empire, this entire region was thriving. The soldiers and hunters shown in the pictograms lived their lives in the middle of a shifting kaleidoscope of empires that were beginning to define the reach of the new Muslim religion.

The petroglyph-makers ceased their activities before the early thirteenth century. At that time, in a swift and probably extremely bloody series of events, all the easternmost of the Turkic empires were conquered and absorbed by Genghis Khan and his Mongol armies.

Vivid representations of animals appear in abundance in all the pictographs that survive from that long period of intensive human occupation. Clearly, the animals must have been numerous enough to be hunted easily across the open country.

Genghis Khan and his warriors changed this relatively stable landscape dramatically. To keep his armies in fighting trim, the Great Khan organized them into vast group hunts. Each army would form a giant ring. As the ring tightened, it swept all the large animals trapped within it into a single struggling, panicked mass. Then the Khan himself would ride into the ring and slaughter the animals until he was sated with killing. His lieutenants continued the slaughter, followed by the cavalry. The killing would sometimes go on for days, unless the Great Khan mercifully stopped it.

But that abundant world was long gone when I visited. In a full day of driving across the exceedingly bleak and stony landscape harboring the petroglyphs, we glimpsed a total of three gazelles, which fled from our Range Rover with astonishing speed. We assumed that these animals must have been subsisting on grasses growing at higher elevations, because there was hardly a trace of green to be found in the whole series of vast empty valley floors that we were traversing.

Of the thirty-one large animal species, including carnivores, that the Mongol hordes could have driven before them, four are now critically endangered, two are endangered, five are vulnerable, and seven are near-threatened. Clearly none are now so abundant that they would be worth sending hungry armies across the landscape to chase them.

The Khan's great slaughters came near the end of a long period of human exploitation of this high-altitude landscape. This exploitation took place in a region that, because of its low rainfall and temperature extremes, must have been subject to repeated stresses even without the added burden imposed by the arrival of humans.

That ecosystem has now been pushed to a tipping point. It no longer functions in the way that ecosystems are supposed to. Most of the large predators that kept the grazing animals' numbers in check have disappeared, along with most of the grazing animals themselves. The grasslands that supported the grazers have succumbed to drought and to overgrazing by domestic animals. Now most of the domesticated animals that had belonged to the bands of herders have also disappeared.

Can this ecosystem recover? Probably not by itself. Ecosystems need organisms, for otherwise they are simply deserts. At the present time, this ancient ecosystem retains only a scattering of its former richness. Might it be feasible to add some factor or factors that could help in restoring its richness—and if so, what?

Hanging On By Their Fingernails

Even when ecosystems are pushed to, and past, their tipping points, they can retain a surprising degree of resourcefulness. Recent metagenomic studies of extreme environments have revealed an extensive pool of microorganisms and their viruses. Even these microorganisms, however, are hanging on by their (figurative) fingernails. A recent study of the exceedingly dry soil of the Atacama Desert of northern Chile, carried out by an international group of ecologists, illustrates this [113]. It also suggests why these organisms have managed to persist.

The Atacama is an extensive desert occupying a large area of northern Chile. It lies between the Andes and the Pacific coast and is the driest place on Earth. This makes it a challenge to survive in, but its extreme atmospheric dryness provides an ideal spot from which we can observe the entire universe.

In 2020, my wife Liz and I visited a microwave observatory in this super-dry zone. Named POLARBEAR, for POLARization of the Background Radiation, it is situated on the slopes of the dormant volcano Cerro Toco at an altitude of more than 5,000 m. From this magnificent vantage, it overlooks the vast Atacama Desert far below.

POLARBEAR's detector is capable of seeing through the desiccated atmosphere and measuring details in the distribution of microwaves from the

Big Bang. It will soon be replaced by a more powerful detector able to find further clues to the rapid cosmic expansion of the early universe.

The unique dryness of this region allows the most distant reaches of the universe to be explored. But it has a huge impact on the organisms that inhabit it. The desert that spreads out far below the observatory appears to be largely lifeless, except for a few isolated canyons where a little water flows. Indeed, one area, known as the Mountains of the Moon, fully lives up to its title.

The ecologists' microbiological survey of the desert's soils found that all of the sites sampled have an unusually low biomass of microorganisms. There are only a few hundred bacteria per gram of soil, compared with numbers in most soils that range from 100 million to ten billion per gram.

It is not yet possible to carry out single-cell genomic studies on such sparse populations. Indeed, even extracting enough DNA from such minuscule samples in order to perform limited metagenome analyses is a challenge. Nonetheless, although these tiny pooled samples of DNA cannot yet give us direct information about which viruses are preying on which bacteria, they do provide ample indirect evidence for such highly specialized host-parasite interactions.

The microbiome study found that the sparse populations of bacteria in the soil fell into four different groups—far fewer than would be expected in soil samples from a less extreme desert. But it also found evidence of the presence of more than 100 different groups of viruses, mostly bacteriophages and mostly new to science.

Many of the viruses carry bacterium genes that had been inserted into their chromosomes, showing that these viruses are playing a role in the horizontal transfer of genes between bacterial species. Some of the bacterial and virus strains share sequences that are known to be involved in bacterial resistance to viruses. The sharing of these resistance sequences shows that host-parasite interactions between bacteria and viruses are still taking place even in such a ghostly remnant of an ecosystem.

The authors of the study suggest that the Atacama microbiomes may be able to evolve further through these bacterium-virus interactions. Thus, this minimally entangled bank may retain the ability to carry out—at least to some extent—Woesian genetic-annealing processes. Even when ecosystems are pushed to the edge, it seems, they may retain some ability to evolve in unexpected ways.

Because of the small numbers of microorganism species involved, these options seem to be sadly limited in the Atacama. The smaller the numbers in a population, the fewer new mutations will appear in it during a given interval of time. Evolution is largely a numbers game, where more mutations mean more possibilities. By this measure, the Atacama does not fare well.

This ecosystem may have additional evolutionary potential, originating from another source. When Liz and I arrived in the town of San Pedro de Atacama, at the beginning of our stay, we were startled to find that the area had just been flooded by a severe rainstorm. This almost unheard-of event had briefly turned the town's little-used gutters into rivers.

We had only a few days in the area, which meant that we had to leave before any effects of the downpour appeared. But records show that even the super-dry Atacama, like Death Valley far to its north, can generate sudden abundant blooms of wildflowers that spring from a "seed bank" of dormant seeds.

Many plant seeds can survive easily for a century or two under such dry conditions. Some can retain their ability to sprout for much longer than that. The oldest authenticated surviving seed was a 2,000-year-old seed of a date palm, discovered in Herod's Palace at Masada in Israel. The tree that was grown from it is still thriving.

Although the Atacama ecosystem's evolutionary options are severely limited, it should still be able to adapt if rainfall should increase in the future.

Restoring Ecosystems from the Bottom Up

The Atacama has been dry for millennia, yet it retains some shreds of evolutionary potential. Other less stressed ecosystems, given the opportunity, might fight back more effectively.

In 1985, shortly after China opened to the outside world, my family and I had the chance to travel to remote parts of the country's west. We got as far as Dunhuang, an ancient town surrounded by great sand dunes on the edge of the Gobi Desert. This pilgrimage site is famous for its many cliff-face caves filled with thousands of depictions of the Buddha.

To get there, we followed the ancient Silk Road west from Lanzhou, the grim and smoky capital of Gansu Province. On the way, we visited Labrang, the largest Tibetan Buddhist monastery in China. The monastery had been founded in 1709, when Labrang was still part of Tibet. It developed into a religious and medical teaching center, and preserved a vast library of ancient Tibetan religious and medical scrolls that were still stored for safekeeping in the monastery's attic. In its prime, the monastery was home to 4,000 monks.

In the early part of the twentieth century, the monastery was caught in the vicious tag-ends of centuries-long conflicts between the Han Chinese and the Muslim groups that lived along China's northern border. After the end of the Second World War, and following China's communist takeover, the Cultural Revolution caused further upheavals.

At the time of our visit, extensive damage to the monastery's buildings by the Red Guards of the Cultural Revolution had only recently been repaired. Most of the people living in the town that surrounded the monastery were still Tibetans. Some of them, in traditional costumes, prostrated themselves repeatedly as they made their prescribed circuits around the main temple building with its hundreds of prayer wheels.

On the final leg of our journey to Dunhuang, we passed the great fort that marks the westernmost end of the rebuilt Ming part of the Great Wall. History dominated our minds, but we were also struck by how barren the area was. We were skirting the southern edge of a remarkable geological formation, the great Loess Plateau.

This grim region, made up of immense hills of wind-blown dust, is webbed by tributaries of the Yellow River. The hills have been built up, over twenty-five million years, by dust carried on the wind storms that swept across the Gobi Desert to the north.

Such dust is not unique to this plateau. Loess dust originates in many desert areas around the world, and the wind can carry it half a world away. But this Chinese plateau is easily the largest deposit of loess dust on the planet.

The landscape that confronted us as we looked out over the undulating plateau was bleak and largely lifeless. Loess is made up of tiny jagged fragments of eroded sand grains, worn down to be so small that they can easily be picked up by the wind. Nothing can grow on loess soil in the absence of water, but if there is water, it can easily support plant life.

To the north of the Loess Plateau, between it and the Gobi of Mongolia, lies the cold, relatively dry Mu Us Desert, covering an area of almost 50,000 km² — larger than Switzerland.

The Mu Us, like the Altai Mountains of western Mongolia to its north that I would visit almost thirty years later, has been inhabited for thousands of years by nomads who hunted game and raised herds of horses and flocks of goats on the land. It was a relatively friendly environment. The lowland areas receive almost half a meter of rain a year, far more than most deserts. In the past, such conditions had been sufficient to support extensive grasslands and even some forested areas. But even before humans arrived, the Mu Us was subject to severe seasonal climatic extremes. And it was always swept by the ferocious winds carrying those clouds of loess dust from the Gobi.

Mu Us means "unpalatable water" in Mongolian. Tribal groups have fought over its diminishing resources for centuries. As the area began to dry up, it rapidly lost vegetation cover, revealing those underlying beds of tiny jagged sand particles.

Slowly, and then in the twentieth century more swiftly, large parts of the Mu Us Desert were transformed into such shifting dunes. The remaining arable land gradually lost its productivity.

In a remarkable memoir published in 2014, Xiao Yinong tells of taking part in massive efforts that began in the mid-1980s to try to stabilize the advancing dunes [114]. He tells how, soon after his arrival, he was caught in a "black sand-storm" so fierce that it carried off and buried full-sized sheep from a nearby village. The jagged loess dust blasted away every trace of paint from the side of the bus in which Yinong had been riding, leaving only bare metal.

Yinong recounts stories of local tribespeople who tried to reverse this over-whelming wind-driven desertification. A village woman, Yu Yuzhen, single-handedly planted thousands of tree saplings and recruited many others to the effort. She is a local heroine, and was nominated for the Nobel Peace Prize.

Alas, only 15 percent of the millions of trees that were planted by the army battalions, and by the many cadres of workers and volunteers like Xiao Yinong, have survived. Conservation workers gradually learned that the ever-changing dunes could not be stabilized rapidly enough using trees alone. Ground cover, which was able to spread quickly and stabilize the shifting landscape, was needed.

Qingfu Liu, along with forestry colleagues in Guizhou and Inner Mongolia and collaborators from around the world, is contributing essential new infor-mation to the Mu Us restoration project [115, 116]. This information is showing that some aspects of the project have been remarkably successful. The results also highlight the many things we do not yet understand about how to restore ecosystems, and hint at possible future directions.

As part of the restoration project, over a period of more than thirty years, seeds were dropped on different parts of the desert from low-flying planes. Because the region is vast, most of the target sites were seeded only once. Each of these one-time seedings was in effect a mini-experiment, because at each of the target sites, the period of time that had elapsed between the aerial seeding and the analysis of its effects was different. The seeding itself was the only inter-vention that was made in these lonely pieces of desert.

Various combinations of seeds were tried, but soon the restorers settled on three that seemed to produce the best results. Throughout much of the experiment, most of the seeds dropped were from two desert leguminous *Hedysarum* plants and a relative of daisies in the genus *Artemisia*. These species are now commonly used throughout China to try to restore desert areas.

The *Hedysarum*, though not native to the area, can increase in size quickly and can grow buds on the belowground parts of their stems. These buds send shoots into the open air to become new plants. And, because *Hedysarum* are legumes, the roots of their seedlings emit chemical signals to attract bacteria that are able to colonize specialized nodules on the roots. Inside the nodules, the plants provide the bacteria with the anoxic environment that they need in order to convert chemically neutral nitrogen gas in the atmosphere to molecules of soluble ammonia that both the bacteria and the plants can utilize.

Their surveys found that the more time that had elapsed since the initial seeding, the greater was the amount of ground cover. The species diversity of these new plant communities also increased. Soon, most of the plants that contributed to this renaissance were not the plants that had grown from the original seeding, but were native to the local desert. The most likely source of this ecological renaissance was native plants that had survived in tiny refugia scattered among the dunes, though seed banks that had been dormant in the soil are likely to have contributed. The plant communities that became established were similar to those in nearby less disturbed areas of the desert.

A decade earlier, the study might have stopped there. But the researchers were able to go further. Metagenomic analyses measured the abundance and overall species diversity of bacteria and fungi in the soil. The numbers and diversity of both also increased dramatically with increasing time since the seedings.

The original seed shower from the airplanes must have included a few bacteria and fungal spores that were clinging to the seeds. There may also have been a few bacteria still clinging to life in the dunes' loess soil. The subsequent increases in microbiome diversity might have been triggered solely by the new ground cover provided by the alien introduced plants as their growing root systems provided soil stability. But it seems likely that there were additional sources of microbial diversity, probably from windblown spores and from "microbiome banks" of spores that are analogous to seed banks and are present even in desert soils. It has yet to be determined how different the microbiomes in the different restored areas will become.

The aerial seeding program and other restoration programs have already had an impact. In many areas, dunes have ceased to spread and sandstorms have dropped in frequency. Some areas that were formerly desert can now support limited grazing and agriculture. But the higher and drier western regions of the desert, after initially showing some improvement, have largely reverted to dunes. The battle for desert restoration is far from won.

Qingfu Liu agrees with me that more carefully targeted seedings, including spores of bacteria and fungi native to desert regions, should be carried out. One essential component of the microbiome, the nitrogen-fixing bacteria that are so important to legume survival, might be added in different quantities to determine whether increases in their number increase the survival and spread of the ground cover. Much experimentation will be needed to determine a safe and effective mix of both bacteria and soil fungi, just as much had been needed to determine the most effective mix of seeds.

Other ecosystems will pose very different problems. But all the restoration efforts should concentrate on restoring the ability of the damaged ecosystems to function well over a range of environments, and to retain and increase their ability to adapt. Success is likely to depend on increases in ecosystem diversity at all levels.

Dozens of studies are probing this interdependence. Guizhou Province, in southwest China near the northern border of Vietnam, is home to many indigenous peoples. Its spectacular landscape, with jutting limestone pillars and extensive cave formations, reflects a long history during which much of the region was covered in shallow seas that supported extensive coral reefs. These landscapes, common throughout the world, are called *karst*, an ancient term that has descended into many forms in many languages and that means "stony."

These areas normally support rich ecosystems. But in large parts of Guizhou, over-intensive farming has eroded away much of the protective soil cover, revealing a kind of tropical desert of bare rock. The deserts persist in spite of the warm climate and abundant rainfall.

Cao et al. [117] found that one essential feature of successful restoration projects in the karst landscapes was the presence of some unusual early colonists of such disturbed areas: the mosses. These plants, and their relatives the tiny liverworts, are bryophytes, even more ancient than the ferns. Together with other associated microflora, they can gradually form a bryophyte crust.

As these bryophyte crusts mature, they become thicker and are able to retain moisture and accumulate soil even on bare rock. Species of moss that can suck up extra water by building up a high concentration of the amino acid proline inside their cells are the most likely to survive. A succession of larger plants, and eventually trees, is able to follow.

Cao et al. followed the details of bryophyte crusts as they matured, and found that the species composition at every level, from bacteria to fungi to mosses, underwent large changes. Different species of fungi and bacteria were associated with each of the moss species, suggesting strong, highly species-dependent mutualistic interactions.

Can the maturation rate of bryophyte crusts be accelerated by appropriate aerial seeding? Detailed studies such as those by Cao et al. may provide clues about how to accomplish this, and how to reverse most effectively the desertification caused by overuse of the land. The mix of species involved, and the nature of the damage that must be repaired, is very different from the situation in the Mu Us Desert. But the overall principle is the same. It is not just the obvious species that have to be restored, but the entire suite of interactions among the species at every level of complexity.

Close examination of the entire suite of organisms in many ecosystems reveals adaptations that depend on previously unsuspected cooperation between species. Desert speargrass rhizosheaths provide a striking example.

These structures take the form of sheaths in the soil around plant roots. Rhizosheaths were first described, at the end of the nineteenth century, in Egyptian desert grass populations. They consist of soil or sand particles, held together by a sticky extracellular polymer that is produced by the many species of bacteria associated with the plants' root hairs. The sheaths, especially well-developed in the root systems of African desert grasses, accumulate water from the surrounding soil or sand, and slow its evaporative loss. The roots and root hairs form a remarkable mutualism with the bacteria that can even build structures when the plant is growing in sand—something that is freely available in the North African desert.

Ramona Marasco of Saudi Arabia's King Abdullah University and her colleagues studied these plant-bacterium interactions and their effect on the maintenance of Sahara Desert grass ecosystems [118]. They took samples of the bacterium populations from the sheaths of endemic speargrass communities that were growing on Tunisian sand dunes.

The sheaths were sampled through the course of an entire year. On these dunes soil temperature changes from summer to winter by as much as 25°C, and humidity by more than 30 percent. Although the microbial community that lived in the sand outside the sheaths changed dramatically in overall species composition from one season to the next, the bacteria associated with the rhizosheaths themselves did not change. Clearly, the plant-microbe mutualism remained stable in spite of large fluctuations in the physical environment and in the rest of the microbial communities. Because of the water storage provided by the rhizosheaths, the grasses that built them were able to survive long periods of severe drought.

The stability of the interactions that produce the rhizosheaths persists over time. Under these extreme desert conditions, the rhizosheaths are essential to the survival of the organisms that take part in them. Any of their component species that evolves in a way that adversely affects its ability to take part in these collaborative interactions is sentencing itself, if not to extinction, at the very least to a marginal existence beyond the protective boundaries of the sheaths themselves.

Most of the Sahara itself is unlikely to be restorable to any kind of stable complex ecosystem unless we succeed in controlling the weather and increasing the rainfall. But the Sahara is already a desert. Parts of the Sahel, the great band of near-desert that extends across much of Africa immediately south of the Sahara, often receive appreciable rainfall. Large regions can support farming and grazing.

One and a half million people in nine countries live in the Sahel, and the population has almost doubled in the last twenty years. But the great Sahara Desert to the north threatens to sweep these precarious ecosystems away.

Estimates of the rate at which the Sahara is advancing into the Sahel vary, but the situation is not helped by the farmers and herders who are pumping out deep reservoirs of old aquifer water left over from the last ice age. This lowers the Sahel's water tables and increases the rate of desertification. Repeated droughts have already triggered local wars, and refugees from these conflicts are crowding into small areas that are able to support far fewer people sustainably.

Successful restoration projects may relieve the pressure on this highly endangered set of ecosystems. But implementation of such projects presents a huge challenge. The Sahel's governments have come together to try to emulate China's huge tree-planting effort and plant billions of trees across the entire region. The project, called the Great Green Wall, sounds as if the idea might have originated in China, but it had multiple origins. In 1950, the British explorer Richard St. Barbe Baker had suggested a "Green Front" to stop the desert's spread.

After more than a decade, the Great Green Wall is only 4 percent of the way to its goal. It has been slowed by political instability and warfare, and by the widespread cutting of the newly planted trees for conversion to charcoal.

Planting trees did not work in the Mu Us, while planting less ambitious ground cover often managed to stabilize the shifting dunes. In the Sahel, it may be that planting of grasses is a better solution for this vast region than the Great Green Wall. Grasses and their rhizosheaths can stabilize the soil and help to retain rainwater, while also supporting grazing animals. Understanding the details of the ecosystems involved, such as the insights provided by Ramona Marasco and her co-workers about plant roots and their associated microbiomes, may help to open up new ways to seed and restore the area. A recent survey of restoration projects throughout Africa by Catherine Parr and associates concluded that forest restoration is often inappropriate and can be damaging, though they did not mention the likelihood that the microbial ecosystems of forests will be poorly adapted to areas better suited to grasslands [149].

These examples show that even ecosystems that are teetering on the very edge of survival can retain a surprising amount of diversity, especially at the microbiome level. And earlier we saw that microbial diversity has facilitated rapid evolutionary change and has sometimes led to remarkable breakthroughs in adaptive ability.

There is no doubt that careful utilization of all these evolutionary and ecological resources in the design of restoration projects can increase their chance of success. This can in turn lead to greater cooperation among farmers, landowners, and governments, all of whom would be able to see immediate benefits in pulling ecosystems back from their tipping points.

How Evolutionary Cauldrons Keep Simmering

In 2016, I visited a coral reef restoration project on a fringing reef just off the south coast of the island of Bohol in the central Philippines. The reefs in the area were largely intact and rich with life. There were more large green sea turtles swimming about in the shallows than I had ever seen in one place, and the deeper reefs were especially spectacular. I remember spending so much time with one multicolored and endearing family of ornate ghost pipefish that I nearly ran out of air.

Some of the reefs close to shore had been severely damaged by local fishermen, who until recently had been using dynamite. When sticks of dynamite are dropped on a reef, they stun or kill thousands of fish. Most of the fish sink to the bottom. Fishermen scoop up the few that float to the surface, and move on to the next reef.

The explosions leave behind smashed corals, along with great cracks in the reefs' ancient foundations. The cracks weaken the reefs' entire structure, so that they may break up during the next typhoon and allow waves to sweep unimpeded toward land. The waves can also smash broken chunks of coral into the living reefs that remain, causing further damage.

Dive clubs and dive resorts on the island have led the fight to ban dynamite fishing. On Bohol, as in many other parts of the Philippines, fishermen who have been shown the magical world that lies beneath the ocean's surface are now among the most vocal and effective defenders of the reefs.

Some restoration projects on these damaged reefs are being attempted, but they face enormous obstacles. I swam out to one, located on a great plateau of bare coral rock that sat in the shallows like the underwater foundation of some great long-vanished building. A grid pattern had been established by a local dive resort, small holes had been drilled at the grid's intersections, and short pieces of staghorn coral had been planted in them.

Some of the pieces of coral had died, but others had survived on this open space for as much as three years. During that time, some but not all of these pieces had increased in length by one or two centimeters. The overall picture was not good. This well-meaning effort was clearly having an extremely depressing outcome.

These corals were, of course, growing in an utterly alien environment. No other corals surrounded them, which meant that their microbiomes must have been reduced in numbers and complexity.

As we saw earlier, reefs were looked on with dread by the sailors in fragile wooden ships like Cook's *Endeavour*. They presented an ever-shifting and ever-resilient danger. Compared with the almost undetectable growth that I saw in the Bohol project, corals in the past seem to have been able to recover swiftly from early navies' dredging operations.

In 2008, I had dived on the German Channel that connects the lagoon of the island of Palau to the sea. Dredged and blasted in 1911 to accommodate large vessels, it is now filling up with exuberant coral growth. The channel is huge and deep and the corals still have a long way to go, but from their substantial size, it is clear that they have been growing far more quickly than the little isolated bits of coral that I saw on the bare reef in Bohol.

We think of damage to coral reefs by humans as having begun only since we began to warm the planet, but it actually extends far back in time. We can trace this history because, as we have seen, corals leave records behind in the form of strata built up from the skeletons of their dead ancestors. One detailed paper, from Richard Norris' group at the Scripps Institution of Oceanography, examined coral growth rates around what is now the harbor at Almirante Bay on the Caribbean coast of Panama [119].

They found that coral growth rates have slowed over the last 3,000 years, decreasing most markedly during the last millennium. During the same period, the densities of parrotfish teeth that were embedded in the old coral layers diminished in direct proportion with the slowing growth, suggesting that intensive fishing by the region's indigenous peoples might have been responsible for both the parrotfish decline and the accompanying coral slowdown. But the data do not rule out other possible causes that might have caused both parrotfish declines and a slowing of reef growth.

Tropical reefs may be the most diverse ecosystems on the planet. Forest Rohwer, a coral virologist at San Diego State University, has estimated that the average coral reef supports ten times as many species, from large to small, as an average rainforest of comparable size.

Some reefs are able to evolve to meet new challenges. We now have detailed data from around the world to back this up.

Stuart Sandin at Scripps Institute of Oceanography has initiated a wide-ranging "Hundred Island Project" to quantify coral adaptation as it is happening. He and his co-workers obtained private funding from the Moore Family Foundation and the Scripps family in order to examine reefs on a hundred islands scattered through the Caribbean, the South Pacific, the Coral Triangle, and the Indian Ocean. The idea was to measure, in a wide range of habitats, detailed aspects of reef growth over time. The project, now almost completed, has done exactly that.

The islands that were surveyed range from those that have experienced severe damage, such as many islands in the Caribbean, to those like the remote uninhabited atolls of the Line Island archipelago in the mid-Pacific that are as untouched as any on the planet. As part of the study, divers snapped wide-angle photographs as they repeatedly followed intricate patterns that had been mapped out on one or more of each island's reefs.

These were not ordinary photographs. They were taken with sharp wide-angle lenses, from different depths and from many locations. Every coral head's shape, volume, color, and texture were recorded in detail. All these terabytes of information were then fed into a computer program that could reconstruct the entire reef in three spatial dimensions and over a span of years. This work is ongoing.

The reefs can be reconstructed on a giant computer screen and examined from any direction. An observer can "swim" through the reef, following any path, and observe each coral head from a variety of angles.

These reconstructed reefs seem to be eerily fish-free, but this is an artifact. So many overlapping images were taken that the fish, which were found in some pictures and not in others, could be edited out.

Because the divers visited these reefs yearly over several years, an observer is able to swim through both space and time. Entire episodes of bleaching and recovery, which may have taken years to happen, can be followed over a span of seconds or minutes on the computer screens.

Sped up in this way, the reefs are revealed to be dynamic places. Plate corals and antler corals are especially active, often able to double in size over a year. But even large brain corals stir and seem to grow visibly, and are sometimes caught in the act of squeezing and choking off invading organisms that try to gain a foothold in their crevices.

The reefs also show sudden invasions, especially of plate corals, that replace parts of the reef. Other coral species soon begin to poke up among the invading plates. The activity on these diverse reefs is utterly different from the miserable stasis of the isolated little bits of corals that I saw at Bohol. The resilience of crowded, thriving coral communities stood in vivid contrast to the inability of those lonely fragments to survive and grow without their millions of visible and invisible companions.

Sandin played a part in another project, led by Michael D. Fox of Woods Hole, which examined the evolutionary response of intact reef communities to repeated severe ocean warming and to the bleaching events that followed. The reefs chosen were those that surrounded the remote Phoenix Islands, halfway between Australia and Hawaii [120].

The reefs were hit by severe bleaching events in 2003, 2009, and 2015, each triggered by strong El Niño events. The first El Niño had the most dramatic impact. Live coral cover decreased from almost 50 percent to about 10 percent. Recovery was slow, reaching 25 percent by the time of the second event in 2009.

Strikingly, even though this second event was weaker than the first, its impact was less than would have been expected. The third event, in 2015, was the strongest of the three. But, while coral cover was significantly reduced, it never dropped below 20 percent. The reefs were, it appeared, adapting to these severe events.

The reefs changed substantially during this time. Corals that had been common on the reefs, such as the highly branched *Pocillopora* and the large plate-like *Acropora*, were almost wiped out by the first bleaching event. But they had largely recovered by 2019.

These and other data from the Phoenix Islands show that corals became more resilient to thermal stress after repeated stresses. But how much of this increase in resilience was the result of natural selection acting on the corals themselves, how much resulted from shifts in the types and numbers of interactions between the corals and their symbionts and pathogens, and how much could be traced to genetic changes in the symbionts that may have modified the severity of the bleaching events? These and many other possibilities have yet to be unraveled.

Evidence is growing that microbiomes play an important role in these changes. Pathogenic bacteria increased in numbers in bleached corals in the Gulf of Thailand [121]. Resistance to bleaching in corals of the South China Sea is correlated with increases in the numbers of species of their zooxanthellae symbionts that are heat-tolerant [122].

We do not yet understand all the between-species interactions involved in coral bleaching. Imagine how far we are from understanding what happens to the millions of interacting species in a reef community when an El Niño hits the reef. And just contemplate for a moment how essential it is to our planet's survival that we do learn exactly what is happening.

The Contradictory Nature of Ecosystems

There are times when science flounders. It can get bogged down for decades in a morass of conflicting observations. We encountered such a morass in Chapter 4, as we traced the decades-long attempts during the nineteenth century to understand the laws that govern inheritance. Now, as ecologists and evolutionary biologists follow new clues about ecosystem function, they are trying to understand a perhaps even more challenging problem, that of finding the apparently complex principles that govern ecosystems.

In their search, they have encountered confusing observations that bear some resemblance to the many seemingly conflicting patterns of inheritance that were encountered by those nineteenth-century plant and animal breeders. But they are aided in their search for the principles of ecological diversity by their growing understand of the transmission of genetic characters and the genetic nature of evolutionary changes.

One major difficulty stands in the way, and perhaps masks these elusive ecological principles from our understanding. Ecosystems possess two seemingly contradictory properties.

First, the fossil record shows that ecosystems can often persist over long periods. The ring of reefs that makes up the Enewetak Atoll has lasted for forty million years. The reef's cast of characters has evolved and changed during that time, but not to the extent that the reef has become unrecognizable. And, provided that the physical conditions around this still-sinking island do not undergo marked shifts, there seems to be no reason why these interacting species cannot continue to exploit such a rich and predictable environment far into the future.

Such persistence suggests that mechanisms are operating that stabilize these ecosystems. But, as we have seen, we can immediately rule out any mechanisms fostering stability that bring a halt to further evolutionary change. In each of an ecosystem's evolving species, evolution is unfettered. It is able to proceed pell-mell as conditions change and as new opportunities arise—and they surely will.

This unfettered evolution constitutes the second of those two contradictory properties. Evolution that favors the survival and reproduction of members of a particular species would seem to be incompatible with the apparent stability of ecosystems. A particularly favorable species should sweep through the entire ecosystem, destabilizing it in the process. There are no ecological or evolutionary principles that would rule such a possibility out. And it sometimes really happens, as in the worldwide upheaval that was caused by the appearance of oxygenic photosynthesis.

The results of evolutionary changes in one species can certainly influence what is happening in the gene pools of other species that share the same ecosystem. But those other species' responses are determined by their own unique genetic history, which will also determine their response to changes in all of the many species with which they interact.

As we have seen, we can rule out group selection, which would have to act somehow on all the species in an ecosystem simultaneously in order to preserve balance and diversity unchanged. Instead, the rules of evolution lead us to conclude that, when a kind of stasis does emerge from the evolutionary cauldron of an entangled ecosystem, it is the product of a teetering balance of contradictory and continuously shifting selective forces. This pseudo-stasis can emerge even though contrary forces are acting on members of each species and mutational changes are happening to those species' gene pools *ad libitum*.

Such a balance may account for some controversial findings of possible interactions among trees in forests, mediated by mutualistic soil fungi [123]. Studies have suggested that such interactions may exchange food and other essentials between trees of the same or different species, and may even transmit signals that "notify" nearby trees of the presence of pathogens and herbivores. Such interactions would, if they played a large role, add greatly to the complexity of entangled-web interactions in forest ecosystems.

These fungus-mediated signals were given the memorable name of the "wood-wide web" [124] in a seminal *Nature* article by Thorunn Helgason of the University of Edinburgh and her colleagues. But such effects have proved frustratingly difficult to quantify, or even to distinguish from alternative explanations [125].

Nonetheless, the wood-wide web has beguiled the public's imagination. Some popular accounts have suggested that this web could be a manifestation of a kind of Gaia-like cooperation, a Peaceable Kingdom, among species (Chapter 8).

Alas, nature is not so kind. The web, if it exists, must depend for its existence on competition and natural selection. It certainly can rely on the evolution of mechanisms of cooperation within groups of species that lead the members of the cooperating groups to become better competitors against species outside the group. But enhancement of competitive ability, either of a species or of a group of species, is still the ultimate determiner of success. The lion is not lying down with the lamb, appealing though that image might be.

A wood-wide web could result from selection within the gene pools of fungi associated with tree roots and the tree species themselves for ways in which they interact with each other. Fungi might easily enter into mutualistic interactions with certain tree species, but only if both would benefit in some way from the ability of the fungi to carry nutrients or messages between trees of the same or even of different tree species. The mutualisms would only last as long as they benefited all the participants. But natural selection rules out the possibility that all tree species could participate in a single such web of positive interactions.

In evolutionary terms, such an elaborate interplay of mutualisms and competition is completely kosher. As we will see shortly, my colleagues and I have found signs of interactions between tree species that would be quite compatible with a less benign form of the wood-wide web.

Another possible solution to the contradiction between the evolution of ecosystems and of their component species was proposed in 2001 by my colleague and collaborator, the remarkable ecologist Stephen Hubbell [126]. In order to explain why ecosystems are so rich in species, he simply canceled out the entire problem. He assumed an ecological equivalence among species, such as species of trees in a forest, that are of the same general kind. Each species does not dominate any of the others. Instead, each species is equivalent in overall fitness to all the other species that occupy similar ecological niches. Indeed, not only are the species equivalent in Steve's model, but new species evolve entirely as a result of random mutations and follow the same rules that random mutations do.

In building his model, Steve did not ignore the multitudinous interactions between species of different kinds, and even of the same kind, that we have encountered throughout this book. But he postulated that these interactions were equivalent for all species, so that all species competed with other species that were similar to them to exactly the same extent. He was able to show that, given the assumptions of his model and given further restrictions such as a careful choice of scale, he could predict distributions of different species of tree in a forest that were quite close matches to the observed distributions.

Numerous critics have attacked Steve's model, showing that it does not often hold. Indeed, he cheerfully agrees that it is best considered a "null" model. It can be thought of as equivalent to the neutral gene models that predict the distribution of genetic alleles in gene pools. In neutral gene models, selection does not operate. His neutral species model suggests that selection does operate, but equivalently for all species. Comparing such a world with the real world, he proposes, may yield useful measures of the variation in selection-driven evolutionary change in various ecosystems.

The evidence we have encountered in this book, along with much more, shows that real ecosystems have properties far different from Steve's null model. But we are still faced with that nagging contradiction between the unpredictable results of interspecies competition and cases of apparent ecosystem stability.

Darwin thought of an ecosystem as a slow-motion explosion of evolving, ever-diverging species. But the Enewetak reef has been "exploding," in evolutionary terms, for forty million years. Even the most elaborate slow-motion

explosion in an action movie, which may seem to the viewer to have been drag-
ging on for several reels, does not last that long!

To summarize, ecosystems are made up of a labile, continually changing collection of species that are furiously competing for resources at the same time as
they are preying on, and otherwise exploiting, each other. Pairs of species or
groups of species may also evolve the ability to cooperate, as a result of selection
within the gene pools of each species for genes that facilitate cooperation.

It is astonishing that such a cacophony of competitive and cooperative interactions can often lead to what appears to an outside observer to be long-term
ecosystem stability. But that stability is illusory. Even such "stable" ecosystems,
with their entangled banks of quarrelling and sometimes cooperating species,
are continually generating new layers of interactions that lead to new opportunities, new ecological niches, and sometimes utterly transformative events.
Cases of apparent ecosystem stability are, over the long term, transitory. Even
on low heat, evolutionary cauldrons continue to bubble.

Sampling Evolutionary Cauldrons

In the 1990s, I took an opportunity to pursue a research direction that was very
different from my usual investigations into evolution in the laboratory. On a
visit to Peru in 1994, I spent a couple of weeks in the rainforest lowlands of
Manú National Park. This was the same park that my wife and I would later
visit during our trip down the eastern slopes of the Andes.

Rainforests are teeming, busy places. But Manú's rainforest is exceptional.
I had never before been confronted by such diversity and dynamic change.
Manú was as far from Steve Hubbell's hypothetical collection of neutral species, each one politely refraining from upsetting the apple cart by outcompet-
ing its neighbors, as it was possible to imagine. How could all this diversity be
maintained over time?

I was primed to explore this question because the dust was only just beginning to settle on the great Kimura neutral-vs.-selected-allele battle among
population geneticists. And I knew that there were many similarities between
genes in gene pools and species in ecosystems.

By 1994, many genetic variants similar to the variants in yeast alcohol
dehydrogenase that my group had generated in the laboratory had been shown
to be strongly selected. But this did not rule out the possibility that some

mutational changes in gene pools really are neutral, though perhaps not as many as Kimura had supposed.

I doubted that species equivalent to Kimura's neutral alleles existed. There are too many things going on in ecosystems, at too many levels of organization, for species to remain aloof from natural selection. But what form does that selection take?

In the 1970s, ecologists Daniel Janzen and Joseph Connell had independently suggested a dynamic balance of selective pressures that could lead to the simultaneous maintenance of many plant or animal species in a forest [127, 128]. They built their model on the phenomenon of *negative density-dependent* or *negative frequency-dependent selection*, which is turning out to be widespread in both gene pools and ecosystems.

Ecologists tend to talk about negative *density*-dependence, and population geneticists tend to talk about negative *frequency*-dependence. This is because geneticists are dealing with gene pools that are continually changing in size, so that the *frequency* of an allele in a population is the best way to keep track of how common or rare the allele is. Ecologists are thinking of the *density* of a species in a given part of its range, measured by its local numbers. I will use the two terms when they are appropriate, but they are really describing very similar phenomena.

An allelic variant of a gene in a gene pool, or a particular species that inhabits an ecosystem, may be at an advantage when it is rare but lose that advantage if it becomes common. Both the allele and the species should increase (in frequency or in density) when they are rare. Then, as they become more common and lose their advantage, the increase will slow and stop. An allele might stay at a balance point with other alleles of that gene indefinitely, but only if selective pressures do not change over time. And a species might stay at a similar balance point with other species in an ecosystem, increasing or decreasing in numbers only if the selective pressures that are acting on it or on the ecosystem's other species change.

To bring this rather dry description to life, let us look at an example of negative frequency (and density) dependence that encompasses both gene pools and ecosystems. The result has been a balanced genetic and ecological polymorphism of mind-boggling complexity.

The story begins with an elaborately polymorphic genetic architecture that has evolved in a single species, allowing that species to interact with several other similar species simultaneously. This genetic polymorphism enables the polymorphic species to effectively occupy several ecological niches that might be expected to support different species.

The genetic polymorphism is found in species of *Papilio* swallowtail butterflies that are widespread in Africa and Asia. Predatory birds find these butterflies toothsome, so that there is an advantage for the butterflies to mimic in appearance any nasty-tasting "model" butterfly species in the vicinity. Such interactions are examples of Batesian mimicry, in which the mimicking species gains an advantage by imitating a distasteful model.

The more common the model species, the more numerous the mimic species can safely be. This is because the mimic is sheltered from predation by the presence of a substantial density of the models, which birds have learned through bitter experience to avoid. But the mimics would quickly become rarer if the models became rare. Batesian mimicry is an example of negative frequency-dependence of polymorphic alleles that allow it to mimic the model species, with the additional feature that the frequencies of these polymorphic alleles are dependent on the density in the ecosystem of its protective model species as well as the density of the mimic itself.

This is fairly simple to think about if there is only one mimic and one model species. The mind-boggling aspect of the *Papilio* mimicry is that the females of a single, randomly mating *Papilio* population often have different wing patterns. This enables different females in the *Papilio* population to mimic a different model species, which provides protection if several model species are present.

The males, however, all look alike. Why don't the males take part in this Batesian mimicry? The males of this species are highly mobile, which makes them more likely to dodge bird predators. The females sit still for long periods in order to attract males. This makes them (to mix a zoological metaphor) sitting ducks for the predators. But the genes for the mimicry system are still present (though repressed) in the males, ready to be passed on to their progeny and to be expressed in the wing patterns of their daughters.

Plate 7 shows females of the African species *Papilio dardanus*, one of the *Papilio* species that mimic several model butterfly species. All of the model species (some are shown to the left in the figure) are in different genera, and therefore very genetically divergent, from the *Papilio* butterflies that mimic them (to the right).

This is not because the *Papilio* females can change their appearance like the river god Proteus. Once a female emerges from her chrysalis, she cannot change her spots, so to speak. But, depending on her genetic makeup and that of the male she mates with, although she expresses only one wing pattern, she carries the ability to have offspring that resemble several different model species.

After much challenging but beautiful work by many geneticists, it was found that the *engrailed* gene, which can control a variety of developmental pathways in different organisms, is a "master regulator" of this astounding polymorphism in the African *Papilio* species [129].

Each of these butterflies carries several sets of genes, each of which codes for developmental pathway steps leading to one of the wing patterns that mimic model species. The *engrailed* gene is polymorphic for a number of "switch" alleles that specify which of these pathways will be activated during the female butterfly's development.

If a model species disappears, the *engrailed* allele in the *Papilio* population that turns on that mimicry pattern gene set may fall in frequency and disappear too. Alternatively, that particular *engrailed* allele may be modified by mutation to turn on a developmental pathway that mimics a new model species.

This complex *Papilio* polymorphism has resulted from a dizzying array of selective pressures acting on these butterflies. The set of *engrailed* alleles and the set of *engrailed*-regulated developmental gene pathways that are present in the gene pool of a population of *P. dardanus* must constantly be evolving as the set of model species that live in the same area changes, both in the numbers of each model and in their local density.

This rococo adaptation is not unique. The Asian swallowtail species *Papilio polytes* has evolved a similar mechanism, but in this species, the gene that regulates pattern development is a different master regulatory gene called *doublesex* [130].

The costs to the *Papilio* females that match a rare model may vary. If the total number of nasty-tasting model butterflies is high, the predatory birds may have learned to be cautious about butterflies in general. This would protect all the mimics (as well as the models). And, if the numbers of different model species are comparable to each other, the mimics might all be equally protected. If a mimic's model is rare, and if the birds remain hungry and somewhat incautious, then females that have that mimic pattern would be at a disadvantage.

In spite of these uncertainties, the *Papilio* approach is still an effective defense against predation. If there were ten model species in an area, then in theory ten independently evolved Batesian mimicry mechanisms in each of ten different mimic species could shield all of them from predation. But each of those mimic species would be at far greater risk of local extinction than the single, much larger, mimetically polymorphic *Papilio* population.

Each *Papilio* model pattern is not critical to the survival of this species of butterfly. If one of the mimic species disappears, most of the *Papilio* females will still be protected. All of the *Papilio* mimics share a single gene pool, which has the capability of adjusting allele frequencies or producing new mimetic patterns through further mutations and genetic recombination relatively quickly.

This example shows that balanced polymorphisms in the gene pool of one of the species in an ecosystem can be influenced by the characteristics of several other species that occupy their own sets of continually shifting equilibrium points in the ecosystem. All these species are far from neutral, in the Hubbell sense of being equivalent to similar species, even before we take into account their interactions with the species that provide their food and with the pathogens and parasites that prey on them.

The likelihood that such complex mechanisms for adaption will evolve is greater in more complex and species-rich ecosystems. But, as we have seen, even apparently simple ecosystems can abound with diversity, especially at the level of the microbiome. Even in oceanic ecosystems where the microbiome plays a huge role, ecological complexities will still provide many opportunities to evolve.

Armed with this little evolutionary detour, we can come back to the forests, and to the puzzle of their diversity that tantalized me as I stood on the rich and teeming shore of an oxbow lake in the middle of the Manú forest. Ecologists are finding that negative density-dependence of the kind that helped shape the *Papilio* butterflies forms a part—but by no means the whole—of the solution to this puzzle.

Data from intact complex ecosystems such as a tropical forest or a coral reef have been essential to unraveling this problem. In 1981, Steve Hubbell and his collaborators Robin Foster and Richard Condit took the first steps toward establishing a completely censused tract of trees in an almost untouched area of tropical forest in Panama. The area they chose was in the middle of a large island.

The island had been formed, early in the twentieth century, by rivers that were dammed to make the huge artificial Lake Gatun that now forms part of the Panama Canal. The island's forest is now protected by the surrounding waters, and by its status as a nature reserve.

The island, named Barro Colorado (which translates to the less romantic-sounding "red mud"), is the site of a research station run by the Smithsonian

Institution, making it an ideal spot to study tropical ecosystems. But the task of establishing the plot was gigantic.

Hubbell and his collaborators realized that to obtain meaningful data about the life histories of the trees, they had to look in unprecedented detail at a large patch of forest. They decided to map carefully all the trees in a half square kilometer patch that were greater than 1 cm in diameter at what used to be known as "breast height." (You will be relieved to learn that this cringeworthy term has now been replaced by "diameter at a height of 1.3 m.")

Each tree had to be identified by species, a challenge in itself because some of the species had not yet been described and young trees are often very different in appearance from older members of the same species. Last, but far from least, the entire census would need to be repeated every five years over a span of decades to follow what was happening to the trees over time. All this would have to be accomplished in spite of the ubiquitous chiggers and mosquitoes, and the roaming herds of peccaries with their uncertain tempers.

They recruited a small army of students and volunteers for the task. After the first census, they ended up with data from more than 300,000 trees of more than 300 different species.

This patch of forest has now been censused seven times, and the approach that was taken has helped to transform forest ecology. By choosing these challenging but feasible census criteria, these far-sighted ecologists established a high standard for subsequent forest research. They have also created a database of unparalleled completeness that has made hundreds of new research projects possible.

The Smithsonian and its collaborators have now established more than sixty of these Barro-Colorado-type *forest diversity plots*, encompassing millions of trees, at sites throughout the world that range from the tropics to the subarctic.

After I got back to the US, during a visit to Princeton, I asked Steve whether I could look at the Barro Colorado data in a way that had not been tried. I wanted to see whether I could detect whether the negative density-dependent effects due to pathogens and predators, which Janzen and Connell had predicted, might be acting among trees of the same species to keep these species at balanced equilibrium numbers. He was delighted to collaborate.

The possibility of such Janzen-Connell effects had tended to be discounted by forest ecologists. They pointed out that the Janzen-Connell model seems to predict that trees of a given species should be widely spaced from each other. This is because the trees' offspring should only be able to survive in distant areas that do not harbor those clusters of pathogens and parasites that the model assumes must surround each large parental tree. Contrary to this apparent prediction, the first data from the Barro Colorado and other forest diversity plots showed that most species are actually growing in clumps.

Rick Condit and I thought that the observed clumping, which is primarily a result of the general tendency of trees to disperse most of their seeds nearby, might be less than would be expected if properties of the trees were shuffled at random within species. Perhaps Janzen-Connell effects could still be operating, and could explain any less-than-predicted clumping of small trees around larger ones. Further, the rate at which trees grow and recruit should be higher in more diverse parts of the forest than in less diverse parts, since pathogens too should be more diverse and less harmful to the trees on average in these more diverse areas.

We came up with a way to measure possible Janzen-Connell effects, by dividing the diversity plots up into tiny sections and repeatedly shuffling the properties of each species of tree between sections. We searched for possible density-dependent effects by comparing these randomized forest diversity plots with the original plot, to see whether the real data showed less clumping of small trees and fewer recruits near large trees than would be expected if the observed distributions were entirely a matter of chance.

We found that sixty-seven of the eighty-four commonest species in the Barro Colorado plot showed strong patterns that were consistent with the Janzen-Connell model. But we cautioned that there were other possible explanations for the patterns that we saw.

Soon, however, experimental data from greenhouse experiments and from several of the diversity plots began to confirm directly that there was indeed a negative density-dependent effect of soil pathogens on seedlings of a tree. Data also emerged showing that more pathogens tended to be shared among closely related tree species, suggesting that these Janzen-Connell effects might extend to interactions between species—especially if they are closely related.

A group of us decided to look in more detail at how Janzen-Connell effects might diminish in strength with increasing evolutionary distance between the tree species.

Our group now included forest ecologists Steve Hubbell and Kyle Harms, the statistical ecologist Thorsten Wiegand, Greg Gilbert who had discovered a negative relationship between the evolutionary distance between species and the numbers of different kinds of parasites that they shared, and the husband-and-wife team of Nimal and Savitri Gunatilleke who had established the Sinharaja forest plot at a World Heritage site in the mountains of Sri Lanka.

At both Barro Colorado and Sinharaja, we found that several different significant measures of negative density-dependence tended to diminish in strength with increasing evolutionary distance between the species [131]. But the picture was complicated, and it showed many cases of apparent *positive* density-dependence. In other words, the interactions between groups of species mostly showed indications that higher density lowered the rate of recruitment, the rate of tree growth, and other measures of tree health. But there were some cases in which higher density had a positive rather than a negative effect on the trees.

More recently, I was part of a larger group of forty-nine ecologists from around the world who took this work a step further. We looked in detail at density-dependent effects at sixteen different diversity plots [132]. These included some plots located in temperate zones, far from the exuberance of the tropics. One such plot was the magical forest of Wytham Woods just outside Oxford, untouched since medieval times. Another was the largely coniferous forest of Wind River in the US state of Washington. We found the same pattern in these northern forests as well as in the tropics and subtropics.

I was especially intrigued by the results from Wytham Woods, which I had visited. This forest is dotted with ancient oaks, beeches, and sycamores, interspersed with tiny meadows of wildflowers. Aside from its great trees, it looks like Darwin's entangled bank.

Wytham is as different from the crowded exuberance of wet tropical forests as could be imagined. Nonetheless, in Wytham and in all the other diversity plots, there was lots of evidence for negative density-dependence. Again, however, there were puzzling exceptions that indicated positive rather than negative between-species interactions.

This massive project would not have been possible without the computer skills of Bin Wang of the Guangxi Institute of Botany in Guilin. With his help, we were able to measure between-species effects on growth, recruitment, and mortality in all the forests. But we were always forced by limitations in our approach to group the species in a plot into clusters made up of species that were separated by similar phylogenetic distances.

Bin and I then set out to find ways to track interactions between particular pairs of species. The majority of species were too rare to provide a meaningful amount of data for analysis, but we found that it was possible to look at interactions between pairs of the commoner species.

These investigations are still at an early stage. But Plate 8, using data from Barro Colorado Island, shows a little sample.

The graphs in Plate 8 show a sample of the species-by-species correlations that we could detect among the commonest twenty-five species in the Barro Colorado Island forest diversity plot. Each little square shows the correlation between the growth rates of trees of a given species and the summed volumes of all the trees of another species that are growing near it. Blue indicates a negative correlation and red a positive one, with the intensity of the color indicating how significant the relationship is. The blank squares indicate that there was no significant correlation.

Running from lower left to upper right (golden dotted line) are all the cases in which correlations were examined between trees of a species and surrounding trees of the same species. Along this line, the correlations are often significant. Many (but not all) of the trees of a given species show a negative correlation between the growth rate of a tree and the total volume of nearby trees of the same species, as one would expect if they were competing for similar resources and sharing pathogens. But there are also some positive correlations.

In the other parts of the graph, away from the dotted golden line, things are also complicated. There are many cases of negative correlations, but there are also some positive correlations between species.

Many of the correlations stay the same from one census period to the next. But many others do not. These shifting correlations can change in significance, and even switch from positive to negative or vice versa, over time.

Perhaps we are looking at Janzen-Connell effects, which could contribute to the negative correlations, and other stimulatory rather than inhibitory effects that could produce positive correlations. Could these analyses provide a way to track down clear cases of positive correlations, such as those between trees and their root fungi that would be expected from the elusive wood-wide web? Data from the microbiomes of the soils in some of these forests are now being analyzed and examined using this approach.

These analyses are just beginning. More extensive analyses, along with experiments involving the species that give the strongest signals, will tell us more.

As far as I know, these forest data provide the best currently available window into the real between-species interactions that are taking place in real time in intact real-world ecosystems, and how these are affected by their associated microbiomes. All of this is made possible because the trees in the forests do not move around! Instead, unlike many organisms, they sit still for repeated censuses.

Such new information will help us to understand how complicated ecosystems work and how their survival can be ensured. In the final chapter, I will look at some ways in which we can harness this growing knowledge.

11

Preserving the World's Evolutionary Cauldrons

[I]n the 1990s researchers began studying…unusual regions of *terra preta do Índio*—rich, fertile "Indian dark earth" that anthropologists believe was made by human beings.

Throughout Amazonia, farmers prize *terra preta* for its great productivity; some have worked it for years with minimal fertilization. Among them are the owners of the papaya orchard I visited, who have happily grown crops on their *terra preta* for two decades. More surprising still, the ceramics in the farm's *terra preta* indicate that the soil has retained its nutrients for as much as a millennium.

Charles C. Mann, *1491: New Revelations of the Americas before Columbus* (Vintage, 2006)

The Greatest Challenge

Today, we can track the challenges that are faced by the world's ecosystems more precisely than ever before. The greatest of these challenges result from the activities of the planet's most unusual species, ourselves.

In many parts of the world, we are still exploding in numbers. The most rapid of these increases are fueled by discrimination against women and lack of access to education, and they are canceling out any benefits of improved public health [133]. Population pressure, and the overexploitation and warfare that it has sparked, have begun to do real damage to our planet. But our evolutionary history has also given us the capacity to overcome these problems.

In the summer of 1974, some colleagues and I were returning to Southern California from a trip to the Sierra Nevada mountains. As we headed south, the highway leading to Los Angeles took us down the gentle slope of the Grapevine Canyon.

The flat expanse of Los Angeles awaited us, hidden as it always was in those days by a thick layer of yellow-brown haze. But there was one unusual feature, a thin layer of bright orange on the top of the haze. It was so vivid that it seemed as if the haze had been covered with icing.

In spite of its innocuous appearance, the orange layer was anything but welcoming. When we descended into it I was driving, and the effect was instantaneous. It was as if I had been tear-gassed. I frantically pulled up on the highway's verge, flailing at my streaming, burning eyes and sending my glasses flying.

I gasped and wheezed, my eyes still stinging furiously, until I had recovered enough to continue driving slowly down through the murderous orange layer. The city's regular choking smog seemed like a welcome relief.

The orange "icing" had formed through chemical reactions between the intense sunlight and the smog below. The result was an especially rich brew of ozone, nitric oxide, and other secondary pollutants.

The smog of Los Angeles has been legendary since early in the twentieth century, but it has abated markedly in recent years. No longer does it tear-gas the unwary. The cleanup has resulted from a combination of enforced car and truck inspections, regulation of polluting industries, reformulation of gasoline, and the rapid spread of electric vehicles.

Less toxic smog was everywhere. The Pacific Ocean near my house just north of San Diego was marred for years by a brown smudge along the entire horizon. Now, when I look out at the Pacific in the early morning, there is little if any haze. I can sometimes glimpse Point Conception, well north of Los Angeles and 200 km away to my northwest.

Other infamous sources of pollution, such as the combustible Cuyahoga River and those deadly English pea-soup smogs known as London Particulars, have been abated. But pollution is still omnipresent over huge tracts of the planet. In 1994, on a bright sunny day, I flew from Mumbai to Kolkata across the entire width of India. Throughout the whole flight, in spite of the good weather, I was unable to glimpse the ground below. It was completely sheathed in dense brown smog, produced by the ancient vehicles, polluting industries, and cooking fires of the teeming world below.

Such huge regions of pollution can have immense consequences. My colleague Veerabhadran Ramanathan [134] and his co-workers have shown that such "brown clouds" can absorb enough sunlight to have a strong negative effect on farm productivity.

Not all is gloom. Some important trends are encouraging. More people can learn about the world than ever before. Global basic literacy, estimated at

Plate 1. A succession of trogon species that we encountered during our descent. A: High-altitude masked trogon female. B: Mid-altitude blue-crowned trogon male. C. Low-altitude black-tailed trogon male. D. Andean Cock-of-the-Rock, *Rupicola peruvianus*.

Plate 2. The Orchis Bank in Kent that may have been the inspiration for Darwin's entangled bank.

Plate 3. Changes in reef ecosystems, from before the Cambrian to the present, showing samples of the life and overall structures of tropical reefs and how they have changed. Artist: D.W. Miller.

Plate 4. A diverse reef at Misool, a small island off the western tip of the island of New Guinea.

Plate 5. A leaf-tailed gecko from Madagascar. Its head, with its pinkish eyes, is to the right of the picture.

Plate 6. A color-polymorphic pair of ornate ghost pipefish on a fringing reef off Bohol in the Philippines.

Plate 7. Model distasteful butterfly species (left) and mimetic females of *P. dardanus* (right) in continental Africa. A male and a non-mimetic *P. dardanus* female from Madagascar are shown at the top of the figure. Copyright Bernard D'Abrera and James Mallet (by permission).

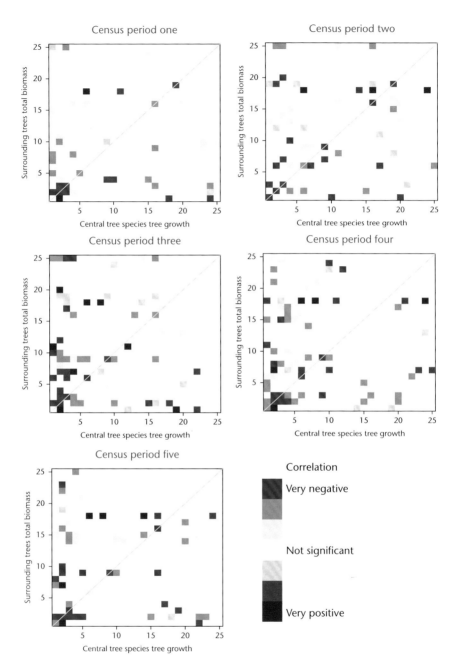

Plate 8. Some of the species-pair interactions between the twenty-five commonest tree species in the Barro Colorado Island forest plot, spanning a total of twenty-five years of periodic censuses. This example shows how the total volume of the surrounding trees of one species is correlated with the growth rates of the central trees of another. Significant negative and significant positive correlations are shown in blue and red, respectively, with the intensity of the color indicating the strength of the correlation. More details in text.

40 percent when I was born at the end of the 1930s, has risen to almost 90 percent. This is especially remarkable because the world's population has increased three and a half times since the moment of my own small contribution.

Worldwide, the expected mean lifespan of our species has almost doubled during my lifetime, primarily through improved public health and education and better ability to transport food. And the overall increase in our population size, which is the result of a combination of changes in both birth rate and life expectancy, has begun to slow markedly. The UN predicts that the planet's human population will begin to decline by the end of this century, though I suspect that this epochal transition will happen more quickly.

Then there is the matter of technology, which, like the amazing genetic mechanisms that have transformed ecosystems throughout the history of life, can be a game-changer. The harnessing of clean and limitless fusion power, which has been called the energy source of the future (and it always will be), has recently been demonstrated to be truly possible. The growing of food is set to be transformed by new methods of low-pollution industrial food manufacture that may take some of the pressure off the world's marginal farmlands. This may give us some extra time to bring the world into balance.

But technology can have vast and unpredictable consequences. For example, our increasing ability to extend human life may result in a speedup of our population's growth and of our planet's decline. If we are suddenly able to live to be 120, will this accelerate the trend toward fewer children? Or will we simply continue down the heedlessly exploitative paths of the past, in order to accommodate our exploding population of Struldbruggian [135] oldsters?

The timing and consequences of all these technological changes (and very likely many others) is unpredictable. All we can be sure of is that they are poised to happen. Unlike the unpredictable changes in gene pools that have led evolution in new directions, these technological changes are predictable because of the growing power of our immense scientific enterprise. And again, unlike evolutionary changes, we can harness these advances and use them to enhance our own—and our planet's—long-term survival. Or not.

Rescuing the world's ecosystems may be made easier by these breakthroughs. But we must continue to reduce our own impact.

Any ecosystem is in danger of being damaged if a species gains a large temporary advantage. But balancing factors will soon appear through the arrival of species from elsewhere or evolutionary changes in the species that are already present. The problem that we face is that we humans have few such checks and balances to limit our own numbers.

The already-fragile ecosystems of the dry western Mongolian valleys were stripped of most of their fauna when Genghis Khan's armies repeatedly swept through them. It might have been possible for these ecosystems to recover from the spectacular depredations of the Mongol armies—if that had been their only challenge. But extensive damage to these lands through overgrazing by horses, cattle, yaks, and goats had begun long before Genghis Khan and continued long after him.

We saw how, in the Mu Us Desert just to the south of these Mongolian valleys, overgrazing and strong desert winds can unleash the destructive power of sand dunes. Advancing dunes can destroy ecosystems at every level of organization, from large predators down to microbes and their viruses. It has been possible to restore some ground cover in the Mu Us, but it is still far from a balanced ecosystem.

The Benefit of Tiny Organisms

We have seen how ecosystems have an amazing ability to reach some degree of balance even under extreme stress, and we have also seen that in most cases it is unclear exactly how this happens. But perhaps, in some cases, it may not be necessary to map the exact road to recovery. In the Mu Us restoration project, seeds that were dropped from the sky were enough to establish a temporary ground cover and to permit the rebuilding of a soil microbiome. It remains to be seen whether additional attempts to replenish depleted microbiomes, followed by the sorting-out of the microorganisms that are added by subsequent rapid natural selection, might improve the speed and effectiveness of ecosystem restoration.

The mix of species making up a microbiome that is successful in a given situation is not pre-ordained. A recent spectacular example of this was provided by Alice Cheng and her colleagues at Stanford [136].

These workers recreated, in the laboratory, a mix of 104 bacterial species that are present in substantial numbers in a typical human microbiome. These were species that could all be grown by the experimenters—provided that they used highly complex mixtures of foods. One of these foods was "chopped meat" media, for example, which is far less well-defined than the simple media that are used to grow most laboratory bacterial strains.

They found that the mix of species stayed stable in culture flasks during repeated laboratory transfers. The next step was to make the mixes even more complex. They took populations of the original mixture and challenged them further by adding samples from human feces, keeping track of those that were able to invade this original mix and join the community. Then, using sterile techniques, they added this relatively small number of successful species to the original set of species, to produce an augmented mixture.

Finally, they made a giant leap. They took both the original mixture and the augmented mixture, and gave each of them to groups of mice that had previously been treated to be germ-free. In both groups of mice, the mixture of these introduced microbial species quickly stabilized, so that the species proportions changed little over time.

Were these artificially constructed microbiomes really stable? To find out, the experimenters then challenged both sets of mice with fresh human fecal pellets, introducing a blast of new species that could not be cultured in the laboratory. Both sets of artificial microbiomes in the mice largely resisted invasion by these new species. The mice carrying the augmented mixture were only slightly more resistant to the invaders than the ones that had been infected with the original mixture.

It appears, remarkably, that a mix of bacterial species that had a long history of coevolving with humans could do quite well in mice and even repel invaders. But it remains to be seen whether, if these human-microbiome mice were put in with regular mice that carry their regular sets of gut microbes, they would eventually revert to a more mouse-like microbiome.

Strikingly, none of the bacterial species got out of control and swept through the microbiome populations. One of the things that seems to happen in microbiomes is that the mix of species that emerges tends toward stability and a resistance to invaders. These properties, it appears, are preserved even when the host organism is switched.

Can this artificial microbiome do all the things that a regular mouse microbiome, with its long history of coevolution with its host organism, can do? Those tasks would include aiding food digestion, generating needed vitamins and other compounds, and defenses against invading pathogens.

It seems unlikely that these newly minted microbiomes could do all of these things well at first, but it also seems possible that the ability of the new microbiome's evolutionary cauldron to shuffle bits of the gene pools of their species mix would eventually give rise to gene combinations that would rectify any deficiencies. But we do not know how quickly this might happen.

Bacterial viruses and other gene transfer agents are likely to be involved in any such further adaptation that takes place. Dr. Cheng and her co-workers did not look at changes in the virus populations during their experiments, but it is high on their agenda.

When microbiomes are introduced in the course of actual ecosystem restorations, as they were at Mu Us, such questions are sure to arise. Soil microbes have been shown to benefit ecosystem restoration in many settings, from urban parks to wilderness areas to near-desert environments. For example, adding *Streptomyces* bacteria can improve survival of Norway spruce seedlings [137]. But the mechanisms and long-term effects are unknown.

In spite of the uncertainties, and the difficulty of preventing the invasion of contaminating organisms, *biopriming* with mixtures of bacteria and mutualistic fungi is a rapidly growing field. I would suggest that this approach holds a special promise. Seeding poor soils with a variety of microorganisms should quickly adjust to an optimal microbiome population, because microorganisms multiply rapidly, quickly adapt to conditions as they evolve and swap genes. When given the chance, they should quickly produce more complete and more multifunctional microbiomes. They may be able to re-establish a healthy evolutionary cauldron that can act as a source for further adaptations.

Such cauldrons are certainly both present and actively cooking in thriving, species-rich soil microbiomes. They may also play a role in the adaptations of more complex organisms.

You will recall those studies of gene transfers between distantly related bacteria in human microbiomes that were carried out by Smillie and his co-workers. These surveys showed high rates of such transfers in human microbiomes, rates that were even higher than the rates observed in the microbiomes of our domestic animals.

Such observations suggest that evolutionary cauldrons are bubbling briskly in the microbiomes of large organisms such as humans. They may be especially active if the host organisms are, like humans, themselves evolving swiftly [138]. And because all the microbiomes of complex organisms that share an ecosystem are linked to each other through surrounding microbiomes that inhabit bodies of water and soil, this activity exponentially increases the evolutionary possibilities that are inherent in every healthy and thriving ecosystem.

Darwin realized that evolution through natural selection has the ability to alter entire ecosystems. We now know that the evolutionary cauldrons that inevitably emerge from the process of natural selection constitute a great strength and resource of ecosystems. These cauldrons are the chief reason that so many ecosystems have survived and even thrived through all of life's vicissitudes.

And there were times, long before the exactitudes of science, when such cauldrons were successfully supplemented, and perhaps even enhanced, by local peoples over long periods of time. Consider the populous but now-vanished agricultural communities of the Amazon basin.

In late 1541, the conquistador Francisco de Orellana and his small band became separated from a larger expedition into the Amazon rainforest. The main expedition was led by Gonzalo, a half-brother of Peru's conqueror, Francisco Pizarro. The goal was to explore the upper Amazon in a search for (what else?) the City of Gold.

When, during a search for food, Orellana's boats were swept away by strong currents, the expedition had already suffered many deaths and injuries from hostile tribes. The rapids eventually slackened and Orellana's band could reach the river's shore, but by this time they were far downstream and separated from the rest of the expedition by rough country and unknown numbers of hostile tribal people. The boats' crews insisted on continuing down the river instead of returning with food to help the rest of the beleaguered expedition.

Orellana acquiesced, perhaps also tempted by the thought that he might be the first to find the City of Gold. This decision would later lead to severe difficulties for him, because Gonzalo Pizarro, along with a handful of the abandoned main expedition, inconveniently managed to survive and return to his brother's army. They were all infuriated by Orellana's traitorous behavior.

Orellana and his crew took eight months to reach the Amazon's mouth, battling warriors from numerous settlements along the way and finally arriving in August of 1542. The trip was chronicled by the Dominican Gaspar de Carvajal, who was probably its only literate member. Carvajal's account was flawed by his wildly embroidered tales of encounters with female warriors, encounters that would give the Amazon its name. But his observations of the many extensive farming communities that they encountered along the river, and how they tended to increase in size and sophistication toward the Amazon's mouth, have been amply confirmed by subsequent excavations.

At the time of Orellana's accidental voyage of discovery, much of the Amazon basin was very different from the seemingly trackless primary rainforest that confronted later explorers. Estimates of the population of the Amazon basin during Orellana's visit vary widely, ranging up to perhaps five million. These peoples were largely wiped out by the plagues of smallpox, measles, and influenza that the Europeans brought with them.

Charles Mann has performed a remarkable reconstruction of this moment in time in his *1491: New Revelations of the Americas before Columbus* [139].

He traces the many efforts by archeologists to understand how such large concentrations of farmers could have thrived in the Amazon basin, perhaps for thousands of years, in an area that tends to have nutrient-poor soils.

Most rainforest ecosystems are notoriously unsuited for agriculture. Typically, when tropical forests are cleared and burned, the bare ground that is revealed can only support two or three seasons of crops before the land has to be abandoned.

The soil of a rainforest, continually washed by drenching rain, is depleted in nutrients. Most of the organic materials are in the living plants and animals, not in the soil. The soils' microbiomes, which have an abundance of fungi and aerobic bacteria, have evolved to facilitate rapid breakdown and recycling of organic materials.

The pre-Columbian agricultural communities of the Amazon may have found a way to slow this extreme recycling, and generate soils that can hold nutrients for longer. Archaeologists and ecologists are discovering large deposits of rich earth in the regions with traces of the agricultural settlements. These soils were formed from a mix of charcoal and waste from the settlements, and are often filled with broken pottery. The mix, called *terra preta do Índio*, is not predominantly ash, unlike the soils in fields that have been slashed and burned. Instead, it is given a more complex structure by its matrix of charcoal, a partially combusted plant material that retains nutrients.

Many indigenous communities living in these areas now thrive on such soils by continuing to maintain them. They carry out sustainable farming of nutrient-rich crops such as cassava. The study of Amazon microbiomes, both natural and human-influenced, is in its infancy, but it is likely to yield valuable insights into how such robust and durable ecosystems can be created.

Introduced Organisms, Benefit and Peril

In 1958, the pioneering ecologist Charles Elton published *The Ecology of Invasions by Animals and Plants* [140]. In this seminal book, he summarized the ecological disasters that have so often followed the introduction of new organisms, whether accidental or deliberate.

Such introductions can have truly disastrous consequences. Brown tree snakes, native to Australia, were spread to many Pacific islands by human migrations. They first arrived on the Micronesian island of Gaum, far to

Australia's north, in the 1950s. Since that time, they have driven half of the island's twenty-five native bird species to extinction.

Australia is both the source and the victim of such disastrous introductions. Anyone who has driven down a country road in northern Queensland and tried to thread a path through a gauntlet of thousands of gigantic, squatting cane toads has confronted such an ecological nightmare.

Cane toads (*Rhinella marina*, formerly *Bufo marinus*) from Central and South America were imported to many areas of the Caribbean and Pacific in the early twentieth century as a biological pest control. The toads, which poison any predator incautious enough to eat them, are especially effective at controlling invasive beetle infestations in sugar cane.

When the toads were introduced to Puerto Rico in 1920, this early attempt at biological control seemed to work quite well. The apparent success led to many introductions throughout the tropics and also, most disastrously, to Queensland and the adjacent Northern Territory of Australia.

The resulting ecological upheaval has put at risk many native reptiles, especially those that prey on nontoxic native amphibians and are poisoned by the cane toads. The decline of these reptilian predators has caused widespread ecological disturbances.

But the cane toads are not totally dominant. In spite of the gigantic size of the adults, which can weigh up to two kilograms in their original habitats, their tadpoles start their lives small and often fall prey to larger tadpoles of native species. Further, since the cane toads were introduced, native predators have begun to evolve avoidance mechanisms, and some show substantial resistance to the cane toads' toxins [141]. The toads' numbers might have been brought under control by this predation, were it not for the fact that they can multiply readily in disturbed and polluted water sources where the native species cannot thrive [142].

Direct competition with native amphibians has also had a variety of results. Some native species in Australia have declined precipitously, but some have actually increased in numbers. The effect of the introduced cane toads has been "rarefaction rather than elimination" [142].

The consequences of other introductions of these toads are also mixed, ranging from the slight to the severe. In the Philippines, where the cane toads have spread to almost all the major islands of this immense archipelago since their introduction in the 1930s, the invasion has had large effects but seems not to have been directly responsible for the extinction of local species [143].

Contemplating these gigantic cane toads, it is hard to avoid the impression that they are a species that only a mother could love. Nonetheless, in their native range in South and Central America, they are responsible for keeping the populations of many insect species in check and are a food source for many predators that are unfazed by their toxins. And, in spite of their less than endearing appearance, the introduced toads have not yet caused the ecological disasters that had been widely feared.

Diverse and thriving ecosystems are remarkably resistant to introductions from outside. As we have seen, these evolutionary cauldrons are made up of thousands of species, each of which has the resources in its gene pool to evolve mechanisms rapidly for resisting invaders. And the invader itself is immediately subject to its own evolutionary pressures.

I am not being a Pollyanna about this. All ecosystems have their limits, and in this book, we have seen many examples of how our own species has an unusual ability to transcend these limits and destroy entire living communities. But healthy ecosystems are remarkably resilient to less disastrous invaders, and have survived such invasions uncounted times in the past.

In 2004, ecologists James H. Brown and Dov Sax proposed a heterodox idea, suggesting that it is impossible to separate invading species into "good" and "bad" [144]. They pointed out that in California, since the arrival of Europeans, more than 1,000 exotic vascular plant species have been introduced from around the world and become established. These invaders now make up a substantial fraction of the 6,000 vascular plant species in the state. During that time, there have been fewer than thirty confirmed extinctions of vascular plants in California (thirteen global extinctions and fifteen local extinctions of species that survive elsewhere).

Some of these introductions have swept through the state, transforming it. In the 1920s and 1930s, George W. Hendry showed that in the late eighteenth century, native perennial grasses were replaced almost totally by fast-growing annual grasses from the Mediterranean. He could follow this replacement by examining the grasses preserved in adobe bricks from that time [145]. The native grasses, once ubiquitous in California's vast Central Valley, have now been reduced to tiny pockets of survivors. But they have not been driven entirely extinct, in spite of the best efforts by humans to co-opt their ecological niches.

Healthy ecosystems are incredibly resilient. In particular, they have enormous capacity to accommodate and adjust to the arrival of new species. Most of the introduced California species are grown in gardens and fields that have replaced grasslands, but the native species that used to inhabit these areas have held their own (though sometimes only barely). The result throughout California is a vast collection of ecosystems that have (despite much upheaval) managed to accommodate themselves to more than a thousand additional multicellular plant species, along with what are probably uncounted millions of additional species of microbes and viruses.

Brown and Sax argue, as I have done here, that today's ecosystems are still far from reaching an ecological limit to the number of species that they can support. Such examples add convincing evidence to Kaustuv Roy's controversial conclusion that shelled molluscs are still exploding in diversity, just as they have done through much of the history of life. And this brings us back to the central theme of this book.

Stoking the Evolutionary Cauldrons

Darwin laid the groundwork for the idea that ecosystems are evolving rapidly, primarily as a result of the evolution—often in divergent directions—of their component species. We have now tunneled down into ecosystems around the world, revealing unsuspected richness of species diversity and species interactions at every organizational level.

We have explored in detail our growing understanding of how species evolve, and have seen some of the many ways that ecosystems are able to retain and to increase their diversity through competitive and cooperative interactions.

Evolution is not something that happened long ago. It is taking place all around us, often at surprising speed. But now, in addition to the many other challenges that ecosystems have always faced, they must adapt to our own species' long-continued onslaught on their integrity.

These stresses will increase in the future, and will likely take unpredictable paths. If global warming continues unabated, large regions of Asia, Africa, and even the Americas may become uninhabitable to humans. If humans cannot live under such extreme conditions, it seems probable that many other multicellular species will not be able to survive either.

Nonetheless, such extreme environments have appeared and disappeared repeatedly in the past, often severely stressing entire ecosystems or changing them beyond recognition. If you were to travel back 8,000 years in time, and be transported to a point anywhere in what is now the Sahara Desert, you would find a green and fertile land of grasses and patches of forest capable of supporting abundant wildlife. This African Humid Period lasted from 14,000 to 5,000 years ago, and it overlapped the spread of early Mediterranean civilizations [146]. Even as it succumbed to desertification, its last vestiges along Africa's north coast made the rise of the Carthaginian Empire possible. Its primary cause was a very humid climate in southern Europe, which spread winter rainfalls further south than they extend today.

This delightful period followed a much more severe event that resulted from global drying during the last glacial maximum. During that time, the Sahara was more extensive than it is today. Desert conditions extended well south into the Sahel.

If you were to travel further back in time, you would find that this dizzying series of wet and dry oscillations occurred repeatedly, through successive periods of circumpolar glaciation and warming.

Clearly, northern African ecosystems have been repeatedly challenged, coming within a whisker of destruction, then able to reconstitute themselves swiftly as conditions eased. Other ecosystems, such as the Amazon Basin, have also undergone large changes in recent times.

In this book, we have explored the genetic and ecological resources that ecosystems possess, traceable to their cornucopia of species and genetic variability at every level of organization. These resources have shaped their resilience.

Our own challenge will be to understand the wellsprings of these diverse living worlds, so that we can assess them as they change and make sure that we know how to supplement them in ways that will restore and preserve their capacity for further evolutionary change. The urgency of this challenge will depend on whether we can change our own behavior.

If things get really desperate, might it be possible to take a "lifeboat" approach, and attempt to preserve ecosystems in isolated areas for some future time when the planet can welcome them? This has been tried, with mixed results—to say the least.

The largest and most elaborate of these experiments began with the building of a large, multi-part sealed greenhouse in the Arizona desert north of Tucson. This structure, constructed with funding by entrepreneur Ed Bass, was called Biosphere 2 (to distinguish it from Biosphere 1, which is nothing less than the planet Earth that Biosphere 2 was designed to imitate).

The complex was designed thoughtfully and well, with the goal of including small, connected ecosystems of rainforest, tropical reef, mangrove swamp, grassland, and desert, along with a section devoted to farming. The whole complex was designed to be sealed off completely from Biosphere 1. The goal was to see whether a small group of humans could live, survive in, and maintain this closed world.

Many possible difficulties were anticipated. Because the air inside the enclosure could expand and contract as the temperature changed, possibly rupturing the domes, "lungs" were designed in the form of inflatable protruding bladders. A dense band of ginger plants shielded the rainforest from the intense desert light. A wide range of plants, and a limited number of animals, were introduced.

Much has been written about the fraught history of the project, which was beset with a multitude of problems [147]. When I visited in 2000, the habitat was open to the outside world (that is, Biosphere 1), so that tourists could go through it. The wooded areas were thriving, but the "coral reef" looked like a murky, badly neglected swimming pool. The project, though still impressive, had strayed far from its original intent.

The two periods during which people lived in the sealed-off complex, the first spanning two years from September of 1991 to September 1993 and the second during part of 1994, yielded some remarkable results.

Unsurprisingly, after its closures the system quickly became unbalanced in a variety of ways. Oxygen slowly diminished, and had to be supplied from outside. Carbon dioxide skyrocketed. Some insects, especially a local ant that had invaded as the biosphere was being built, quickly multiplied at the expense of others. Pollinating insects, which were essential to the human crew's farming efforts, did not survive for long.

The first human crew had trouble living on what they could raise, but those troubles diminished as they became more inventive. The second crew drew on the mistakes of the first, and managed their food supply more successfully.

The result that I think has the most relevance to ecosystem resilience is what happened to the most complex of the small ecosystems, the rainforest. It was only two-tenths of a hectare in size, and was initially planted with 1,833 plants of 282 different species of New World tropical trees and shrubs.

These were more trees than this little area could support, and it was expected that many species would be lost. Indeed, by 1993, the numbers had fallen to 588 trees of 172 species. The numbers declined further to 528 trees of 130 species by 1996 [148].

How do these numbers compare with the numbers in a real tropical forest?

The number of species currently living in the Barro Colorado forest plot is 320 in an area 50 ha in size. The number of trees (excluding those less than 1 cm in diameter) is 360,000. I calculated the average number of trees and species in a 0.2 ha piece of this real forest plot, and found that the number of trees was 1,472 and the number of species was 132. The tiny rainforest in Biosphere 2 had lost many trees and species, but still had as many species as a similar area of the highly diverse Barro Colorado forest. And the rate of species loss seemed to be slowing. It would be fascinating to know what had happened to this experimental rainforest's microbiome during this period!

Would such a lifeboat ecosystem work? Clearly, it did retain some diversity over five years. An experiment that lasted 500 years might, however, yield a less sanguine result.

On top of that, small numbers are the bane of a healthy ecosystem. It is hard to imagine a lifeboat world of the future surviving if all its pollinating insects had disappeared, as they did in Biosphere 2.

In spite of all the controversies surrounding the Biosphere 2 project, it has provided us with another look at the power of ecosystems to adapt to new challenges. But we can—and must—do better than desperately fill such lifeboats if our planet is to recover from the challenges of our own making.

In this book, we have explored some of the brilliant breakthroughs in science that are giving us glimpses of the way forward. We have as a species seized such moments in the past, and I am serenely confident that we can do so again.

Music, art, science, and literature from around the world have transformed and enriched the way we all live. We are indeed a wonderful species to have created such things! We have learned and understood so much about how the world works, and we have enriched it in so many ways. Surely we are worth preserving, so that we can accomplish even more amazing things in the future. But it is becoming more and more obvious that, if we are to achieve our full potential, we must bear the cost of preserving the rest of the amazing living beings—no matter how small—with which we share our wonderful planet.

We are presented with a stark choice. The evolutionary payoff if we learn to behave altruistically toward the rest of the world will be huge. If we don't, the evolutionary payoff will be zero—both for ourselves and for our planet.

Bibliography

1. Jankowski J.E., Merkord C.L., Rios W.F., Cabrera K.G., Revilla N.S., Silman M.R., Gillman L.N. 2013 The relationship of tropical bird communities to tree species composition and vegetation structure along an Andean elevational gradient. *Journal of Biogeography* **40**(5), 950–62. (doi:10.1111/jbi.12041).

2. Thery M., Larpin D. 1993 Seed dispersal and vegetation dynamics at a cock-of-the-rocks lek in the tropical forest of French-Guiana. *Journal of Tropical Ecology* **9**, 109–16. (doi:10.1017/s0266467400007045).

3. Marting P.R., Kallman N.M., Wcislo W.T., Pratt S.C. 2018 Ant-plant sociometry in the *Azteca-Cecropia* mutualism. *Scientific Reports* **8**(1), 17968. (doi:10.1038/s41598-018-36399-9).

4. Pringle E.G., Akcay E., Raab T.K., Dirzo R., Gordon D.M. 2013 Water stress strengthens mutualism among ants, trees, and scale insects. *PLOS Biology* **11**(11). (doi:10.1371/journal.pbio.1001705).

5. Gutierrez-Valencia J., Chomicki G., Renner S.S. 2017 Recurrent breakdowns of mutualisms with ants in the neotropical ant-plant genus *Cecropia* (Urticaceae). *Molecular Phylogenetics and Evolution* **111**, 196–205. (doi:10.1016/j.ympev.2017.04.009).

6. Wills C. 2010 *The Darwinian Tourist: Viewing the World through Evolutionary Eyes*. Oxford, Oxford University Press.

7. Kent W.J., Baertsch R., Hinrichs A., Miller W., Haussler D. 2003 Evolution's cauldron: duplication, deletion, and rearrangement in the mouse and human genomes. *Proceedings of the National Academy of Sciences of the United States of America* **100**(20), 11484–9. (doi:10.1073/pnas.1932072100).

8. Bell G. 1982 *The Masterpiece of Nature: The Evolution and Genetics of Sexuality*. Berkeley, University of California Press.

9. Brea M., Artabe A., Spalletti L.A. 2008 Ecological reconstruction of a mixed Middle Triassic forest from Argentina. *Alcheringa: An Australasian Journal of Palaeontology* **32**(4), 365–93. (doi:10.1080/03115510802417760).

10. Darwin C., Darwin F. 1987 *The Foundations of the Origin of Species: Two Essays Written in 1842 and 1844*. London, William Pickering.

11. Darwin E. 1818 *Zoonomia*. Pennsylvania, E. Earle.

12. Zirkle C. 1946 The early history of the idea of the inheritance of acquired characters and of pangenesis. *Transactions of the American Philosophical Society* **35**(2), 91–151. (doi:10.2307/1005592).

13. Hughes A.L., Yeager M. 1998 Natural selection at major histocompatibility complex loci of vertebrates. *Annual Review of Genetics* **32**, 415–35. (doi:10.1146/annurev.genet.32.1.415).

14. Haeckel E. 1870 Über Entwickelungsgang u. Aufgabe der Zoologie. *Jenaische Z* **5**, 353–70.

15. Macarthur R.H., Wilson E.O. 1963 Equilibrium-theory of insular zoogeography. *Evolution* **17**(4), 373–87. (doi:10.2307/2407089).

16. Simberloff D.S., Wilson E.O. 1970 Experimental zoogeography of islands—a 2-year record of colonization. *Ecology* **51**(5), 934–7. (doi:10.2307/1933995).

17. Quinzin M.C., Sandoval-Castillo J., Miller J.M., Beheregaray L.B., Russello M.A., Hunter E.A., Gibbs J.P., Tapia W., Villalva F., Caccone A. 2019 Genetically informed captive

breeding of hybrids of an extinct species of Galapagos tortoise. *Conservation Biology* **33**(6), 1404–14. (doi:10.1111/cobi.13319).

18. Gause G.F., Nastukova O.K., Alpatov W.W. 1934 The influence of biologically conditioned media on the growth of a mixed population of *Paramecium caudatum* and *P. aurelia*. *Journal of Animal Ecology* **3**(2), 222–30. (doi:10.2307/1145).

19. Crombie A.C., Imms A.D. 1946 Further experiments on insect competition. *Proceedings of the Royal Society of London Series B—Biological Sciences* **133**(870), 76–109. (doi:10.1098/rspb.1946.0004).

20. Macarthur R. 1960 On the relative abundance of species. *American Naturalist* **94**(874), 25–36. (doi:10.1086/282106).

21. Riordon W.L., Plunkitt G.W. 2001 *Plunkitt of Tammany Hall: A Series of Very Plain Talks on Very Practical Politics, Delivered by ex-Senator George Washington Plunkitt, the Tammany Philosopher, from His Rostrum—the New York County Court House Bootblack Stand*. The Project Gutenberg.

22. Gunz P., Bookstein F.L., Mitteroecker P., Stadlmayr A., Seidler H., Weber G.W. 2009 Early modern human diversity suggests subdivided population structure and a complex out-of-Africa scenario. *Proceedings of the National Academy of Sciences of the United States of America* **106**(15), 6094–8. (doi:10.1073/pnas.0808160106).

23. Higham C. 2013 Hunter-gatherers in southeast Asia: from prehistory to the present. *Human Biology* **85**(1–3), 21-43. (doi:10.3378/027.085.0302).

24. Aubert M., Lebe R., Oktaviana A.A., Tang M., Burhan B., Hamrullah, Jusdi A., Abdullah, Hakim B., Zhao J.X., et al. 2019 Earliest hunting scene in prehistoric art. *Nature* **576**(7787), 442–5. (doi:10.1038/s41586-019-1806-y).

25. Skov L., Macia M.C., Sveinbjornsson G., Mafessoni F., Lucotte E.A., Einarsdottir M.S., Jonsson H., Halldorsson B., Gudbjartsson D.F., Helgason A., et al. 2020 The nature of Neanderthal introgression revealed by 27,566 Icelandic genomes. *Nature* **582**(7810), 78–83. (doi:10.1038/s41586-020-2225-9).

26. Huerta-Sanchez E., Jin X., Asan, Bianba Z., Peter B.M., Vinckenbosch N., Liang Y., Yi X., He M., Somel M., et al. 2014 Altitude adaptation in Tibetans caused by introgression of Denisovan-like DNA. *Nature* **512**(7513), 194–7. (doi:10.1038/nature13408).

27. Jacobs G.S., Hudjashov G., Saag L., Kusuma P., Darusallam C.C., Lawson D.J., Mondal M., Pagani L., Ricaut F.X., Stoneking M., et al. 2019 Multiple deeply divergent Denisovan ancestries in Papuans. *Cell* **177**(4), 1010–21. (doi:10.1016/j.cell.2019.02.035).

28. Turney C.S.M., Bird M.I., Fifield L.K., Roberts R.G., Smith M., Dortch C.E., Grün R., Lawson E., Ayliffe L.K., Miller G.H., et al. 2001 Early human occupation at Devil's Lair, southwestern Australia 50,000 years ago. *Quaternary Research* **55**(1), 3–13. (doi:https://doi.org/10.1006/qres.2000.2195).

29. van der Kaars S., Miller G.H., Turney C.S.M., Cook E.J., Nurnberg D., Schonfeld J., Kershaw A.P., Lehman S.J. 2017 Humans rather than climate the primary cause of Pleistocene megafaunal extinction in Australia. *Nature Communications* **8**. (doi:10.1038/ncomms14142).

30. Woinarski J.C., Burbidge A.A., Harrison P.L. 2015 Ongoing unraveling of a continental fauna: decline and extinction of Australian mammals since European settlement. *Proceedings of the National Academy of Sciences of the United States of America* **112**(15), 4531–40. (doi:10.1073/pnas.1417301112).

31. Wells R.T., Camens A.B. 2018 New skeletal material sheds light on the palaeobiology of the Pleistocene marsupial carnivore, *Thylacoleo carnifex*. *PLOS One* **13**(12). (doi:10.1371/journal.pone.0208020).

32. Wroe S. 2002 A review of terrestrial mammalian and reptilian carnivore ecology in Australian fossil faunas, and factors influencing their diversity: the myth of reptilian domination and its broader ramifications. *Australian Journal of Zoology* **50**(1), 1–24. (doi:10.1071/ZO01053).

33. Perry G.L.W., Wheeler A.B., Wood J.R., Wilmshurst J.M. 2014 A high-precision chronology for the rapid extinction of New Zealand moa (Aves, Dinornithiformes). *Quaternary Science Reviews* **105**, 126–35. (doi:https://doi.org/10.1016/j.quascirev.2014.09.025).

34. Koungoulos L., Fillios M. 2020 Hunting dogs down under? On the Aboriginal use of tame dingoes in dietary game acquisition and its relevance to Australian prehistory. *Journal of Anthropological Archaeology* **58**, 101146. (doi:https://doi.org/10.1016/j.jaa.2020.101146).

35. Saltre F., Rodriguez-Rey M., Brook B.W., Johnson C.N., Turney C.S., Alroy J., Cooper A., Beeton N., Bird M.I., Fordham D.A., et al. 2016 Climate change not to blame for late Quaternary megafauna extinctions in Australia. *Nature Communications* **7**, 10511. (doi:10.1038/ncomms10511).

36. Robins J.H., McLenachan P.A., Phillips M.J., McComish B.J., Matisoo-Smith E., Ross H.A. 2010 Evolutionary relationships and divergence times among the native rats of Australia. *BMC Evolutionary Biology* **10**. (doi:10.1186/1471-2148-10-375).

37. Sacks B.N., Brown S.K., Stephens D., Pedersen N.C., Wu J.T., Berry O. 2013 Y chromosome analysis of dingoes and southeast Asian village dogs suggests a Neolithic continental expansion from southeast Asia followed by multiple Austronesian dispersals. *Molecular Biology and Evolution* **30**(5), 1103–18. (doi:10.1093/molbev/mst027).

38. Tuft K., Legge S., Frank A.S.K., James A.I., May T., Page E., Radford I.J., Woinarski J.C.Z., Fisher A., Lawes M.J., et al. 2021 Cats are a key threatening factor to the survival of local populations of native small mammals in Australia's tropical savannas: evidence from translocation trials with *Rattus tunneyi*. *Wildlife Research* **48**(7), 654–62. (doi:10.1071/WR20193).

39. Burchfield J.D. 1975 *Lord Kelvin and the Age of the Earth*. New York, Science History Publications.

40. During M.A.D., Smit J., Voeten D., Berruyer C., Tafforeau P., Sanchez S., Stein K.H.W., Verdegaal-Warmerdam S.J.A., van der Lubbe J. 2022 The Mesozoic terminated in boreal spring. *Nature* **603**(7899), 91–4. (doi:10.1038/s41586-022-04446-1).

41. Henry T.W., Wardlaw B.R. 1990 *Introduction: Enewetak Atoll and the PEACE Program*. Washington, Department of the Interior, U.S. Geological Survey.

42. Nutman A.P., Bennett V.C., Friend C.R.L., Van Kranendonk M.J., Chivas A.R. 2016 Rapid emergence of life shown by discovery of 3,700-million-year-old microbial structures. *Nature* **537**(7621), 535–8. (doi:10.1038/nature19355).

43. Hanada S., Takaichi S., Matsuura K., Nakamura K. 2002 *Roseiflexus castenholzii* gen. nov., sp nov., a thermophilic, filamentous, photosynthetic bacterium that lacks chlorosomes. *International Journal of Systematic and Evolutionary Microbiology* **52**, 187–93. (doi:10.1099/00207713-52-1-187).

44. Casaburi G., Duscher A.A., Reid R.P., Foster J.S. 2016 Characterization of the stromatolite microbiome from Little Darby Island, the Bahamas using predictive and whole shotgun metagenomic analysis. *Environmental Microbiology* **18**(5), 1452–69. (doi:10.1111/1462-2920.13094).

45. Turner E.C. 2021 Possible poriferan body fossils in early Neoproterozoic microbial reefs. *Nature* **596**(7870), 87–91. (doi:10.1038/s41586-021-03773-z).

46. Shu D. 2008 Cambrian explosion: birth of tree of animals. *Gondwana Research* **14**(1–2), 219–40. (doi:10.1016/j.gr.2007.08.004).

47. Fedonkin M.A., Waggoner B.M. 1997 The late Precambrian fossil *Kimberella* is a mollusc-like bilaterian organism. *Nature* **388**(6645), 868–71. (doi:10.1038/42242).

48. Morris S.C. 2000 The Cambrian "explosion": slow-fuse or megatonnage? *Proceedings of the National Academy of Sciences of the United States of America* **97**(9), 4426–9. (doi:10.1073/pnas.97.9.4426).

49. Jiang L., Zhao M., Shen A., Huang L., Chen D., Cai C. 2022 Pulses of atmosphere oxygenation during the Cambrian radiation of animals. *Earth and Planetary Science Letters* **590**, 117565. (doi:https://doi.org/10.1016/j.epsl.2022.117565).

50. Johnson C.C. 2002 The rise and fall of rudist reefs. *American Scientist* **90**(2), 148–53. (doi:10.1511/2002.2.148).

51. Larson B., Ruiz-Herrero T., Lee S., Kumar S., Mahadevan L., King N. 2020 Biophysical principles of choanoflagellate self-organization. *Proceedings of the National Academy of Sciences of the United States of America* **117**(3), 1303–11. (doi:10.1073/pnas.1909447117).

52. Lipps J.H., Stanley G.D. 2016 Reefs through time: an evolutionary view. In *Coral Reefs at the Crossroads* (eds. Hubbard D.K., Rogers C.S., Lipps J.H., Stanley J.G.D.), pp. 175–96. Dordrecht, Springer Netherlands.

53. Jablonski D., Roy K., Valentine J.W., Price R.M., Anderson P.S. 2003 The impact of the pull of the recent on the history of marine diversity. *Science* **300**(5622), 1133–5. (doi:10.1126/science.1083246).

54. Raup D.M. 1979 Biases in the fossil record of species and genera. *Bulletin of Carnegie Museum of Natural History* **13**, 85–91.

55. Galton F. 1871 Pangenesis. *Nature* **4**(79), 5–6. (doi:10.1038/004005b0).

56. Dobzhansky T.G., Epling C.C. 1944 *Contributions to the Genetics, Taxonomy, and Ecology of Drosophila Pseudoobscura and Its Relatives.* Washington, DC, Carnegie Institute.

57. Dobzhansky T., Pavlovsky O., Spassky B., Willis C.J., Anderson W.W. 1964 Genetics of natural-populations. XXXV. Progress report on genetic changes in populations of *Drosophila-pseudoobscura* in American Southwest. *Evolution* **18**(2), 164–176. (doi:10.2307/2406389).

58. Fuller Z.L., Haynes G.D., Richards S., Schaeffer S.W. 2017 Genomics of natural populations: evolutionary forces that establish and maintain gene arrangements in *Drosophila pseudoobscura. Molecular Ecology* **26**(23), 6539–62. (doi:10.1111/mec.14381).

59. Franklin R.E., Gosling R.G. 1953 Evidence for 2-chain helix in crystalline structure of sodium deoxyribonucleate. *Nature* **172**(4369), 156–7. (doi:10.1038/172156a0).

60. Watson J.D., Crick F.H.C. 1953 Molecular structure of nucleic acids: a structure for deoxyribose nucleic acid. *Nature* **171**(4356), 737–8. (doi:10.1038/171737a0).

61. Kreitman M. 1983 Nucleotide polymorphism at the alcohol-dehydrogenase locus of *Drosophila melanogaster. Nature* **304**(5925), 412–17. (doi:10.1038/304412a0).

62. Spencer P.S., Siller E., Anderson J.F., Barral J.M. 2012 Silent substitutions predictably alter translation elongation rates and protein folding efficiencies. *Journal of Molecular Biology* **422**(3), 328–35. (doi:10.1016/j.jmb.2012.06.010).

63. Reshef D.N., Reshef Y.A., Finucane H.K., Grossman S.R., McVean G., Turnbaugh P.J., Lander E.S., Mitzenmacher M., Sabeti P.C. 2011 Detecting novel associations in large data sets. *Science* **334**(6062), 1518–24. (doi:10.1126/science.1205438).

64. Maynard-Smith J., Haigh J. 1974 Hitch-hiking effect of a favorable gene. *Genetics Research* **23**(1), 23–35. (doi:10.1017/S0016672300014634).

65. Kjaer K.H., Winther Pedersen M., De Sanctis B., De Cahsan B., Korneliussen T.S., Michelsen C.S., Sand K.K., Jelavic S., Ruter A.H., Schmidt A.M.A., et al. 2022 A 2-million-year-old ecosystem in Greenland uncovered by environmental DNA. *Nature* **612**(7939), 283–91. (doi:10.1038/s41586-022-05453-y).

66. Hutchinson G.E. 1961 The paradox of the plankton. *American Naturalist* **95**(882), 137–45. (doi:10.1086/282171).

67. Handelsman J., Liles M., Mann D., Riesenfeld C., Goodman R.M. 2002 Cloning the metagenome: culture-independent access to the diversity and functions of the uncultivated microbial world. In *Functional Microbial Genomics* (eds. Wren B., Dorrell N.), pp. 241–55. London, Academic Press.

68. Browne H.P., Forster S.C., Anonye B.O., Kumar N., Neville B.A., Stares M.D., Goulding D., Lawley T.D. 2016 Culturing of 'unculturable' human microbiota reveals novel taxa and extensive sporulation. *Nature* **533**(7604), 543–6. (doi:10.1038/nature17645).

69. Macosko E.Z., Basu A., Satija R., Nemesh J., Shekhar K., Goldman M., Tirosh I., Bialas A.R., Kamitaki N., Martersteck E.M., et al. 2015 Highly parallel genome-wide expression profiling of individual cells using nanoliter droplets. *Cell* **161**(5), 1202–14. (doi:10.1016/j.cell.2015.05.002).

70. Bankevich A., Nurk S., Antipov D., Gurevich A.A., Dvorkin M., Kulikov A.S., Lesin V.M., Nikolenko S.I., Pham S., Prjibelski A.D., et al. 2012 SPAdes: a new genome assembly algorithm and its applications to single-cell sequencing. *Journal of Computational Biology* **19**(5), 455–77. (doi:10.1089/cmb.2012.0021).

71. Klunk J., Vilgalys T.P., Demeure C.E., Cheng X., Shiratori M., Madej J., Beau R., Elli D., Patino M.I., Redfern R., et al. 2022 Evolution of immune genes is associated with the Black Death. *Nature* **611**(7935), 312–19. (doi:10.1038/s41586-022-05349-x).

72. Danovaro R., Dell'Anno A., Corinaldesi C., Magagnini M., Noble R., Tamburini C., Weinbauer M. 2008 Major viral impact on the functioning of benthic deep-sea ecosystems. *Nature* **454**(7208), 1084–U1027. (doi:10.1038/nature07268).

73. Martinez-Hernandez F., Diop A., Garcia-Heredia I., Bobay L.M., Martinez-Garcia M. 2022 Unexpected myriad of co-occurring viral strains and species in one of the most abundant and microdiverse viruses on Earth. *The ISME Journal* **16**(4), 1025–35. (doi:10.1038/s41396-021-01150-2).

74. Brumfield K.D., Usmani M., Chen K.E.M., Gangwar M., Jutla A.S., Huq A., Colwell R.R. 2021 Environmental parameters associated with incidence and transmission of pathogenic *Vibrio* spp. *Environmental Microbiology* **23**(12), 7314–40. (doi:10.1111/1462-2920.15716).

75. Wright A.C., Hill R.T., Johnson J.A., Roghman M.C., Colwell R.R., Morris J.G. 1996 Distribution of *Vibrio vulnificus* in the Chesapeake Bay. *Applied and Environmental Microbiology* **62**(2), 717–24. (doi:10.1128/AEM.62.2.717-724.1996).

76. Vezzulli L., Grande C., Reid P.C., Helaouet P., Edwards M., Hofle M.G., Brettar I., Colwell R.R., Pruzzo C. 2016 Climate influence on *Vibrio* and associated human diseases during the past half-century in the coastal North Atlantic. *Proceedings of the National Academy of Sciences of the United States of America* **113**(34), E5062–71. (doi:10.1073/pnas.1609157113).

77. Yang J., Pu J., Lu S., Bai X.N., Wu Y.F., Jin D., Cheng Y.P., Zhang G., Zhu W.T., Luo X.L., et al. 2020 Species-level analysis of human gut microbiota with metataxonomics. *Frontiers in Microbiology* **11**. (doi:10.3389/fmicb.2020.02029).

78. Mills D.R., Peterson R.L., Spiegelman S. 1967 An extracellular Darwinian experiment with a self-duplicating nucleic acid molecule. *Proceedings of the National Academy of Sciences of the United States of America* **58**(1), 217–24+. (doi:10.1073/pnas.58.1.217).

79. Mizuuchi R., Furubayashi T., Ichihashi N. 2022 Evolutionary transition from a single RNA replicator to a multiple replicator network. *Nature Communications* **13**(1), 1460. (doi:10.1038/s41467-022-29113-x).

80. Kamiura R., Mizuuchi R., Ichihashi N. 2022 Plausible pathway for a host-parasite molecular replication network to increase its complexity through Darwinian evolution. *PLOS Computational Biology* **18**(12). (doi:10.1371/journal.pcbi.1010709).

81. Kimura M. 1979 Neutral theory of molecular evolution. *Scientific American* **241**(5), 98–100. (doi:10.1038/scientificamerican1179-98).

82. Wills C. 1973 In defense of naive pan-selectionism. *American Naturalist* **107**(953), 23–34. (doi:10.1086/282814).

83. Wills C., Jornvall H. 1979 Amino-acid substitutions in 2 functional mutants of yeast alcohol-dehydrogenase. *Nature* **279**(5715), 734–6. (doi:10.1038/279734a0).

84. Wills C. 1976 Production of yeast alcohol-dehydrogenase isoenzymes by selection. *Nature* **261**(5555), 26–9. (doi:10.1038/261026a0).

85. McDonald M.J., Rice D.P., Desai M.M. 2016 Sex speeds adaptation by altering the dynamics of molecular evolution. *Nature* **531**(7593), 233–6. (doi:10.1038/nature17143).

86. Leu J.Y., Chang S.L., Chao J.C., Woods L.C., McDonald M.J. 2020 Sex alters molecular evolution in diploid experimental populations of *S. cerevisiae*. *Nature Ecology & Evolution* **4**(3), 453–60. (doi:10.1038/s41559-020-1101-1).

87. Smillie C.S., Smith M.B., Friedman J., Cordero O.X., David L.A., Alm E.J. 2011 Ecology drives a global network of gene exchange connecting the human microbiome. *Nature* **480**(7376), 241–4. (doi:10.1038/nature10571).

88. Garud N.R., Good B.H., Hallatschek O., Pollard K.S. 2019 Evolutionary dynamics of bacteria in the gut microbiome within and across hosts. *PLOS Biology* **17**(1). (doi:10.1371/journal.pbio.3000102).

89. Stevenson C., Hall J.P., Harrison E., Wood A., Brockhurst M.A. 2017 Gene mobility promotes the spread of resistance in bacterial populations. *The ISME Journal* **11**(8), 1930–2. (doi:10.1038/ismej.2017.42).

90. Nguyen A.N.T., Woods L.C., Gorrell R., Ramanan S., Kwok T., McDonald M.J. 2022 Recombination resolves the cost of horizontal gene transfer in experimental populations of *Helicobacter pylori*. *Proceedings of the National Academy of Sciences of the United States of America* **119**(12), e2119010119. (doi:10.1073/pnas.2119010119).

91. Woods L.C., Gorrell R.J., Taylor F., Connallon T., Kwok T., McDonald M.J. 2020 Horizontal gene transfer potentiates adaptation by reducing selective constraints on the spread of genetic variation. *Proceedings of the National Academy of Sciences of the United States of America* **117**(43), 26868–75. (doi:10.1073/pnas.2005331117).

92. Voolstra C.R., Ziegler M. 2020 Adapting with microbial help: microbiome flexibility facilitates rapid responses to environmental change. *Bioessays* **42**(7), e2000004. (doi:10.1002/bies.202000004).

93. Wills C. 1989 *The Wisdom of the Genes: New Pathways in Evolution*. New York, Basic Books.

94. Liu S.F., Lu H.I., Chi W.L., Liu G.H., Kuo H.C. 2023 Sniffer dogs diagnose lung cancer by recognition of exhaled gases: using breathing target samples to train dogs has a higher diagnostic rate than using lung cancer tissue samples or urine samples. *Cancers* **15**(4). (doi:10.3390/cancers15041234).

95. Marchalonis J.J., Schluter S.F., Bernstein R.M., Hohman V.S. 1998 Antibodies of sharks: revolution and evolution. *Immunological Reviews* **166**, 103–22. (doi:10.1111/j.1600-065X.1998.tb01256.x).

96. Das S., Rast J.P., Li J., Kadota M., Donald J.A., Kuraku S., Hirano M., Cooper M.D. 2021 Evolution of variable lymphocyte receptor B antibody loci in jawless vertebrates. *Proceedings of the National Academy of Sciences of the United States of America* **118**(50). (doi:10.1073/pnas.2116522118).

97. Carmona L.M., Fugmann S.D., Schatz D.G. 2016 Collaboration of RAG2 with RAG1-like proteins during the evolution of V(D)J recombination. *Genes & Development* **30**(8), 909–17. (doi:10.1101/gad.278432.116).

98. Sabeti P.C., Walsh E., Schaffner S.F., Varilly P., Fry B., Hutcheson H.B., Cullen M., Mikkelsen T.S., Roy J., Patterson N., et al. 2005 The case for selection at CCR5-Delta 32. *PLOS Biology* **3**(11), 1963–9. (doi:10.1371/journal.pbio.0030378).

99. Libert F., Cochaux P., Beckman G., Samson M., Aksenova M., Cao A., Czeizel A., Claustres M., De la Rúa C., Ferrari M., et al. 1998 The delta CCR5 mutation conferring protection against HIV-1 in Caucasian populations has a single and recent origin in Northeastern Europe. *Human Molecular Genetics* **7**, 399–406. (doi:10.1093/hmg/7.3.399).

100. Rodriguez-Mora S., De Witid F., Garcia-Perez J., Bermejo M., Lopez-Huertas M.R., Mateos E., Marti P., Rocha S., Vigon L., Christ F., et al. 2019 The mutation of Transportin 3 gene that causes limb girdle muscular dystrophy 1F induces protection against HIV-1 infection. *PLOS Pathogens* **15**(8). (doi:10.1371/journal.ppat.1007958).

101. Hall A.R., Scanlan P.D., Morgan A.D., Buckling A. 2011 Host-parasite coevolutionary arms races give way to fluctuating selection. *Ecology Letters* **14**(7), 635–42. (doi:10.1111/j.1461-0248.2011.01624.x).

102. Nowak M.A., Tarnita C.E., Wilson E.O. 2010 The evolution of eusociality. *Nature* **466**(7310), 1057–62. (doi:10.1038/nature09205).

103. Wills C., Bada J. 2000 *The Spark of Life: Darwin and the Primeval Soup*. Cambridge, Mass, Perseus Pub.

104. Woese C. 1998 The universal ancestor. *Proceedings of the National Academy of Sciences of the United States of America* **95**(12), 6854–9. (doi:10.1073/pnas.95.12.6854).

105. Imachi H., Nobu M.K., Nakahara N., Morono Y., Ogawara M., Takaki Y., Takano Y., Uematsu K., Ikuta T., Ito M., et al. 2020 Isolation of an archaeon at the prokaryote-eukaryote interface. *Nature* **577**(7791), 519–25. (doi:10.1038/s41586-019-1916-6).

106. Spang A., Saw J.H., Jorgensen S.L., Zaremba-Niedzwiedzka K., Martijn J., Lind A.E., van Eijk R., Schleper C., Guy L., Ettema T.J.G. 2015 Complex archaea that bridge the gap between prokaryotes and eukaryotes. *Nature* **521**(7551), 173–9. (doi:10.1038/nature14447).

107. Ha P.T., Lindemann S.R., Shi L., Dohnalkova A.C., Fredrickson J.K., Madigan M.T., Beyenal H. 2017 Syntrophic anaerobic photosynthesis via direct interspecies electron transfer. *Nature Communications* **8**(1), 13924. (doi:10.1038/ncomms13924).

108. He H.P., Wu X., Zhu J.X., Lin M., Lv Y., Xian H.Y., Yang Y.P., Lin X.J., Li S., Li Y.L., et al. 2023 A mineral-based origin of Earth's initial hydrogen peroxide and molecular oxygen. *Proceedings of the National Academy of Sciences of the United States of America* **120**(13). (doi:10.1073/pnas.2221984120).

109. Schirrmeister B.E., Gugger M., Donoghue P.C.J. 2015 Cyanobacteria and the great oxidation event: evidence from genes and fossils. *Palaeontology* **58**(5), 769–85. (doi:10.1111/pala.12178).

110. Sperling E.A., Boag T.H., Duncan M.I., Endriga C.R., Marquez J.A., Mills D.B., Monarrez P.M., Sclafani J.A., Stockey R.G., Payne J.L. 2022 Breathless through time: oxygen and animals across Earth's history. *Biological Bulletin* **243**(2), 184–206. (doi:10.1086/721754).

111. Tripathi B.M., Edwards D.P., Mendes L.W., Kim M., Dong K., Kim H., Adams J.M. 2016 The impact of tropical forest logging and oil palm agriculture on the soil microbiome. *Molecular Ecology* **25**(10), 2244–57. (doi:10.1111/mec.13620).

112. Li X., Jousset A., de Boer W., Carrión V.J., Zhang T., Wang X., Kuramae E.E. 2019 Legacy of land use history determines reprogramming of plant physiology by soil microbiome. *The ISME Journal* **13**(3), 738–51. (doi:10.1038/s41396-018-0300-0).

113. Fletcher L.E., Conley C.A., Valdivia-Silva J.E., Perez-Montano S., Condori-Apaza R., Kovacs G.T., Glavin D.P., McKay C.P. 2011 Determination of low bacterial concentrations in hyperarid Atacama soils: comparison of biochemical and microscopy methods with real-time quantitative PCR. *Canadian Journal of Microbiology* **57**(11), 953–63. (doi:10.1139/w11-091).

114. Yinong X. 2014 *Looking for Mu Us*, China Intercontinental Press.

115. Liu Q., Zhang Q., Jarvie S., Yan Y., Han P., Liu T., Guo K., Ren L., Yue K., Wu H., et al. 2021 Ecosystem restoration through aerial seeding: interacting plant–soil microbiome effects on soil multifunctionality. *Land Degradation & Development* **32**(18), 5334–47. (doi:10.1002/ldr.4112).

116. Liu Q., Zhao Y., Zhang X., Buyantuev A., Niu J., Wang X. 2018 Spatiotemporal patterns of desertification dynamics and desertification effects on ecosystem services in the Mu Us Desert in China. *Sustainability* **10**(3). (doi:10.3390/su10030589).

117. Cao W., Xiong Y., Zhao D., Tan H., Qu J. 2020 Bryophytes and the symbiotic microorganisms, the pioneers of vegetation restoration in karst rocky desertification areas in southwestern China. *Applied Microbiology and Biotechnology* **104**(2), 873–91. (doi:10.1007/s00253-019-10235-0).

118. Marasco R., Fusi M., Mosqueira M., Booth J.M., Rossi F., Cardinale M., Michoud G., Rolli E., Mugnai G., Vergani L., et al. 2022 Rhizosheath-root system changes exopolysaccharide content but stabilizes bacterial community across contrasting seasons in a desert environment. *Environmental Microbiome* **17**(1), 14. (doi:10.1186/s40793-022-00407-3).

119. Cramer K.L., O'Dea A., Clark T.R., Zhao J.X., Norris R.D. 2017 Prehistorical and historical declines in Caribbean coral reef accretion rates driven by loss of parrotfish. *Nature Communications* **8**, 14160. (doi:10.1038/ncomms14160).

120. Fox M.D., Cohen A.L., Rotjan R.D., Mangubhai S., Sandin S.A., Smith J.E., Thorrold S.R., Dissly L., Mollica N.R., Obura D. 2021 Increasing coral reef resilience through successive marine heatwaves. *Geophysical Research Letters* **48**(17). (doi:10.1029/2021gl094128).

121. Kusdianto H., Kullapanich C., Palasuk M., Jandang S., Pattaragulwanit K., Ouazzani J., Chavanich S., Viyakarn V., Somboonna N. 2021 Microbiomes of healthy and bleached corals during a 2016 thermal bleaching event in the Upper Gulf of Thailand. *Frontiers in Marine Science* **8**. (doi:10.3389/fmars.2021.643962).

122. Qin Z.J., Yu K.F., Chen S.C., Chen B., Liang J.Y., Yao Q.C., Yu X.P., Liao Z.H., Deng C.Q., Liang Y.T. 2021 Microbiome of juvenile corals in the outer reef slope and lagoon of the South China Sea: insight into coral acclimatization to extreme thermal environments. *Environmental Microbiology* **23**(8), 4389–404. (doi:10.1111/1462-2920.15624).

123. Simard S.W., Perry D.A., Jones M.D., Myrold D.D., Durall D.M., Molina R. 1997 Net transfer of carbon between ectomycorrhizal tree species in the field. *Nature* **388**(6642), 579–82. (doi:10.1038/41557).

124. Helgason T., Daniell T.J., Husband R., Fitter A.H., Young J.P.W. 1998 Ploughing up the wood-wide web? *Nature* **394**(6692), 431–1. (doi:10.1038/28764).

125. Karst J., Hoeksema J.D., Jones M.D., Turkington R. 2011 Parsing the roles of abiotic and biotic factors in Douglas-fir seedling growth. *Pedobiologia* **54**(5-6), 273–80. (doi:10.1016/j.pedobi.2011.05.002).

126. Hubbell S.P. 2001 *The Unified Neutral Theory of Biodiversity and Biogeography*. Princeton, Princeton University Press.

127. Connell J.H. 1971 On the role of natural enemies in preventing competitive exclusion in some marine animals and in rain forest trees. In *Dynamics of Populations* (eds. Den Boer P.J., Gradwell G.), pp. 298–312. New York, PUDOC.

128. Janzen D.H. 1970 Herbivores and the number of tree species in tropical forests. *American Naturalist* **104**, 501–29.

129. Timmermans M., Baxter S.W., Clark R., Heckel D.G., Vogel H., Collins S., Papanicolaou A., Fukova I., Joron M., Thompson M.J., et al. 2014 Comparative genomics of the mimicry switch in *Papilio dardanus*. *Proceedings of the Royal Society of London Series B—Biological Sciences* **281**(1787). (doi:10.1098/rspb.2014.0465).

130. Kunte K., Zhang W., Tenger-Trolander A., Palmer D.H., Martin A., Reed R.D., Mullen S.P., Kronforst M.R. 2014 Doublesex is a mimicry supergene. *Nature* **507**(7491), 229–32. (doi:10.1038/nature13112).

131. Wills C., Harms K.E., Wiegand T., Punchi-Manage R., Gilbert G.S., Erickson D., Kress W.J., Hubbell S.P., Gunatilleke C.V.S., Gunatilleke I. 2016 Persistence of neighborhood demographic influences over long phylogenetic distances may help drive post-speciation adaptation in tropical forests. *PLOS One* **11**(6). (doi:10.1371/journal.pone.0156913).

132. Wills C., Wang B., Fang S., Wang Y.Q., Jin Y., Lutz J., Thompson J., Harms K.E., Pulla S., Pasion B., et al. 2021 Interactions between all pairs of neighboring trees in 16 forests worldwide reveal details of unique ecological processes in each forest, and provide windows into their evolutionary histories. *PLOS Computational Biology* **17**(4). (doi:10.1371/journal. pcbi.1008853).

133. Leung T.Y., Sharma P., Adithipyangkul P., Hosie P. 2020 Gender equity and public health outcomes: the COVID-19 experience. *Journal of Business Research* **116**, 193–8. (doi:https:// doi.org/10.1016/j.jbusres.2020.05.031).

134. Ramanathan V., Chung C., Kim D., Bettge T., Buja L., Kiehl J.T., Washington W.M., Fu Q., Sikka D.R., Wild M. 2005 Atmospheric brown clouds: impacts on South Asian climate and hydrological cycle. *Proceedings of the National Academy of Sciences of the United States of America* **102**(15), 5326–33. (doi:10.1073/pnas.0500656102).

135. Swift J. 1960 *Gulliver's Travels*. London, Oxford University Press.

136. Cheng A.G., Ho P.Y., Aranda-Diaz A., Jain S., Yu F.B., Meng X., Wang M., Iakiviak M., Nagashima K., Zhao A., et al. 2022 Design, construction, and in vivo augmentation of a complex gut microbiome. *Cell* **185**(19), 3617–36 e3619. (doi:10.1016/j.cell.2022.08.003).

137. Lehr N.A., Schrey S.D., Hampp R., Tarkka M.T. 2008 Root inoculation with a forest soil streptomycete leads to locally and systemically increased resistance against phytopathogens in Norway spruce. *New Phytologist* **177**(4), 965–76. (doi:10.1111/j.1469-8137.2007.02322.x).

138. Mathieson I., Lazaridis I., Rohland N., Mallick S., Patterson N., Roodenberg S.A., Harney E., Stewardson K., Fernandes D., Novak M., et al. 2015 Genome-wide patterns of selection in 230 ancient Eurasians. *Nature* **528**(7583), 499–503. (doi:10.1038/nature16152).

139. Mann C.C. 2005 *1491: New Revelations of the Americas before Columbus*. 1st ed. New York, Knopf.

140. Elton C.S. 1958 *The Ecology of Invasions by Animals and Plants*. London, Methuen.

141. Tingley R., Phillips B.L., Letnic M., Brown G.P., Shine R., Baird S.J.E. 2013 Identifying optimal barriers to halt the invasion of cane toads *Rhinella marina* in arid Australia. *Journal of Applied Ecology* **50**(1), 129–37. (doi:10.1111/1365-2664.12021).

142. Taylor A., McCallum H.I., Watson G., Grigg G.C. 2017 Impact of cane toads on a community of Australian native frogs, determined by 10 years of automated identification and logging of calling behaviour. *Journal of Applied Ecology* **54**(6), 2000–10. (doi:10.1111/1365-2664.12859).

143. Harvey J.A., Ambavane P., Williamson M., Diesmos A. 2022 Evaluating the effects of the invasive cane toad (*Rhinella marina*) on island biodiversity, focusing on the Philippines. *Pacific Conservation Biology* **28**(3), 199–210. (doi:10.1071/PC21012).

144. Brown J.H., Sax D.F. 2004 An essay on some topics concerning invasive species. *Austral Ecology* **29**(5), 530–6. (doi:10.1111/j.1442-9993.2004.01340.x).

145. Hendry G.W. 1931 The adobe brick as a historical source: reporting further studies in adobe brick analysis. *Agricultural History* **5**(3), 110–27.

146. Cheddadi R., Carre M., Nourelbait M., Francois L., Rhoujjati A., Manay R., Ochoa D., Schefuss E. 2021 Early Holocene greening of the Sahara requires Mediterranean winter rainfall. *Proceedings of the National Academy of Sciences of the United States of America* **118**(23). (doi:10.1073/pnas.2024898118).

147. Nelson M. 2018 *Pushing Our Limits: Insights from Biosphere 2*. Tucson, The University of Arizona Press.

148. Leigh L.S., Burgess T., Marino B.D.V., Wei Y.D. 1999 Tropical rainforest biome of Biosphere 2: structure, composition and results of the first 2 years of operation. *Ecological Engineering* **13**(1–4), 65–93. (doi:10.1016/S0925-8574(98)00092-5).

149. Parr, C.L., Te Beest, M., Stevens, N. 2024 Conflation of reforestation with restoration is widespread. *Science* **383**(6684), 698–701. (doi:10.1126/science.adj0899)

Index

For the benefit of digital users, indexed terms that span two pages (e.g., 52–53) may, on occasion, appear on only one of those pages.